T0142207

Lecture Notes in Statistics **173**

Edited by P. Bickel, P. Diggle, S. Fienberg, K. Krickeberg,
I. Olkin, N. Wermuth, and S. Zeger

Springer
New York
Berlin
Heidelberg
Hong Kong
London
Milan
Paris
Tokyo

Jesper Møller (Editor)

Spatial Statistics and Computational Methods

Jesper Møller
Department of Mathematical Sciences
Aalborg University
Fredrik Bajers Vej 7G
DK-9220 Aalborg
Denmark
jm@math.auc.dk

Library of Congress Cataloging-in-Publication Data
Spatial statistics and computational methods / editor, Jesper Møller
p. cm. — (Lecture notes in statistics ; 173)
Includes bibliographical references and index.
ISBN 0-387-00136-0 9 (pbk. : acid-free paper)
1. Geography — Statistical methods. 2. Spatial analysis (Statiscs) I. Møller, Jesper. II. Lecture notes in statistics (Springer-Verlag) ; v. 173.
G70.3 .S57 2002
910'.01'5195—dc21 2002042667

ISBN 0-387-00136-0 Printed on acid-free paper.

Printed in the United States of America.

9 8 7 6 5 4 3 2 1 SPIN 10900314

Typesetting: Pages created by the authors using a Springer TEX macro package.

www.springer-ny.com

Springer-Verlag New York Berlin Heidelberg
A member of BertelsmannSpringer Science+Business Media GmbH

Preface

Spatial statistics is concerned with the analysis of spatially indexed data, and some of the earliest applications in statistics of Markov Chain Monte Carlo (MCMC) methods are within spatial statistics. Spatial statistics and MCMC techniques have each undergone major developments in the last decade. Also, these two areas are mutually reinforcing, because MCMC methods are often necessary for the practical implementation of spatial statistical inference, whilst new spatial stochastic models in turn motivate the development of improved MCMC algorithms.

This volume gives an introduction to topics of current interest in spatial and computational statistics, which should be accessible to postgraduate students as well as to experienced statistical researchers. The volume covers many of the scientific areas of the European Union's TMR network "Statistics and Computational Methods for the Analysis of Spatial Data". It is partly based on the course-material for the "TMR and MaPhySto Summer School on Spatial Statistics and Computational Methods", held at Aalborg University, Denmark, August 19–22, 2001.

The volume consists of the following four chapters.

Chapter 1 (by Petros Dellaportas and Gareth O. Roberts) gives a tutorial on MCMC methods, the computational methodology which is essential for virtually all the complex spatial models to be considered in subsequent chapters. It gives an introduction to Metropolis-Hastings algorithms beginning with the Gibbs sampler, and moving progressively through more complex methods. A short section on basic MCMC theory is given, although no specialist knowledge of Markov chains is assumed. Furthermore, a discussion on practical issues related to the implementation of algorithms is given. This includes issues of how to sample efficiently from the relevant conditional distributions, some ideas about when algorithms converge slowly and how to avoid this, a brief introduction to output analysis, and a description of methods for assessing Monte Carlo error from MCMC samples. Chapter 1 also gives a very simple spatial example which serves to illustrate the earlier material in this chapter. Although analytically tractable, this example shares many of the characteristic features of the more realistic spatial models considered in Chapter 2, and provides a simple testing ground for basic techniques. The chapter contains an appendix on MCMC methods for trans-dimensional spaces, focusing in particular on model determination using MCMC methods. The chapter is not intended to be a self-contained introduction to MCMC, but it does provide sufficient background for the substantial uses of MCMC in Chapters 2–4.

Chapter 2 (by Peter J. Diggle, Paulo J. Ribeiro Jr., and Ole F. Christensen) concerns model-based geostatistics. The term geostatistics identifies the part of spatial statistics which is concerned with the study of a spatial phenomenon which exists throughout a continuous spatial region but is only observed at a finite sample of locations. The subject has its origins in problems connected with estimation of ore reserves in the mining industry, but rapidly found applications in other subject-areas. Most of the early development of geostatistical methods took place largely independently of "mainstream" spatial statistics. Chapter 2 introduces the reader to the model-based approach to geostatistics, by which is meant the application of general statistical principles to the formulation of explicit stochastic models for geostatistical data, and to inference within a declared class of models. The chapter emphasises the use of hierarchically specified stochastic models in which the observable quantities are modelled conditionally on a latent Gaussian spatial stochastic process. Bayesian inference then provides an integrated approach to problems of parameter estimation and spatial prediction.

Chapter 3 (by Merrilee A. Hurn, Oddvar K. Husby, and Håvard Rue) begins with the aims of image analysis — what is the goal of a statistical treatment of the data which constitute a digital image? Aims can be quantitative or qualitative, and this may affect the modelling and the inference. Throughout the chapter a Bayesian viewpoint is taken. The chapter explores some of the classes of prior models available, beginning with Markov random field (MRF) models for binary and categorical data. The Hammersley-Clifford theorem is discussed, and the Ising model (and some generalisations) are considered in detail. The chapter demonstrates different types of noise process acting on binary images, and MCMC simulation techniques both for the prior and the posterior model. Point estimation for images is considered via the construction of loss functions. Inference for nuisance parameters is also considered, including a fully Bayesian approach. The chapter then turns to grey-level images, discussing how the modelling might need to change, and giving some details of Gaussian MRFs. As a final class of image types, the chapter considers high level imaging, looking at polygonal templates in a marked point process framework. Finally the chapter works through an example in ultrasound imaging to illustrate many of the ideas.

Chapter 4 (by Jesper Møller and Rasmus P. Waggepetersen) collects recent theoretical advances in simulation-based inference for spatial point processes. Two examples of applications are used throughout the chapter for illustrative purposes. Simple and marked point processes in the d-dimensional Eucledian space as well as a general setting for point processes are introduced without going too much into technical details. The chapter surveys some of the most important properties of Poisson processes and classical non-parametric methods for summary statistics, both in the stationary and nonstationary case. The primary focus of the chapter is on point process models which are analysed using a likelihood or Bayesian approach by means of MCMC methods. Particularly two main classes of models are studied, namely Cox and Markov point process models. Important subclasses of Cox processes such as Neyman-Scott processes and log Gaussian Cox processes are studied in more detail. Also pairwise interaction point processes and more general examples of Markov point processes are given. Moreover, in the different sections dealing with Poisson, Cox, and Markov point processes various algorithms for simulation, including perfect simulation, are discussed.

All the chapters together show how sophisticated spatial statistical and computational methods apply to a range of problems of increasing importance for applications in science and technology. As the four chapters are intimately related several cross-references are given. However, each chapter can be read independently of the others, though the reader with a limited knowledge on MCMC may like to read Chapter 1 before proceeding with any of the Chapters 2–4.

Acknowledgments: The "TMR and MaPhySto Summer School on Spatial Statistics and Computational Methods" was supported by the European Union's TMR network "Statistical and Computational Methods for the Analysis of Spatial Data. ERB-FMRX-CT96-0095" and by the Centre for Mathematical Physics and Stochastics (MaPhySto), funded by a grant from The Danish National Research Foundation. Each chapter has been refereed by an external referee and by a contributor of another chapter: Ole F. Christensen, Arnoldo Frigessi, Peter Guttorp, Merrilee A. Hurn, Antti Penttinen, Hårvard Rue, Håkon Tjelmeland, and Rasmus P. Waagepetersen are acknowledged for their useful comments. Further acknowledgements can be found at the end of each chapter.

Aalborg, August 2002 Jesper Møller

Contents

Contributors

Ole F. Christensen
Department of Mathematics and Statistics
Lancaster University
Lancaster LA1 4YF
UK
Email: o.christensen@lancaster.ac.uk

Petros Dellaportas
Department of Statistics
Athens University of Economics and Business
76 Patission Street
GR-10434 Athens
Greece
Email: petros@aueb.gr

Peter J. Diggle
Department of Mathematics and Statistics
Lancaster University
Lancaster LA1 4YF
UK
Email: p.diggle@lancaster.ac.uk

Merrilee A. Hurn
Department of Mathematics
University of Bristol
Bristol BS8 1TW
UK
Email: M.Hurn@bris.ac.uk

Oddvar K. Husby
Department of Mathematical Sciences
The Norwegian University of Science and Technology
N-7491 Trondheim
Norway
Email: okh@math.ntnu.no

Jesper Møller
Department of Mathematical Sciences
Aalborg University
DK-9220 Aalborg Øst
Denmark
Email: jm@math.auc.dk

Paulo J. Ribeiro Jr.
Departamento de Estatística
Universidade Federal do Paraná
Caixa Postal 19.081
CEP 81.531 - 990 Curitiba, PR.
Brazil
Email: Paulo.Ribeiro@est.ufpr.br

Gareth O. Roberts
Department of Mathematics and Statistics
Lancaster University
Lancaster LA1 4YF
UK
Email: G.O.Roberts@lancaster.ac.uk

Hårvard Rue
Department of Mathematical Sciences
The Norwegian University of Science and Technology
N-7491 Trondheim
Norway
Email: havard.rue@math.ntnu.no

Rasmus P. Waagepetersen
Department of Mathematical Sciences
Aalborg University
DK-9220 Aalborg Øst
Denmark
Email: rw@math.auc.dk

1

An Introduction to MCMC

Petros Dellaportas
Gareth O. Roberts

1.1 MCMC and spatial statistics

Markov chain Monte Carlo (MCMC) algorithms are now widely used in virtually all areas of statistics. In particular, spatial applications featured very prominently in the early development of the methodology (Geman & Geman 1984), and they still provide some of the most challenging problems for MCMC techniques. It is not of a great surprise that most modern introductions to the subject emphasise the flexibility and generality of MCMC. We will do that here too, but we also have more targeted aims in supporting the other chapters in this book covering different areas of spatial statistics.

The problem which MCMC addresses is a familiar one throughout a broad range of diverse disciplines including physics, engineering, computer science and statistics. We are interested in simulating from a probability distribution, call it $\pi(x)$. Although the function $\pi(\cdot)$ can be written in closed form (at least up to a normalisation constant), suppose that we are interested in moments of π which are not computable analytically. Furthermore, direct simulation from π may be difficult or impossible, perhaps because of its high dimensionality, so that simple Monte Carlo evaluation of these moments of π is not possible. MCMC attacks this problem by instead simulating from a Markov chain whose invariant distribution is π.

As we shall see, there are only two basic techniques used in MCMC. One uses an accept/reject mechanism to 'correct' an arbitrary Markov chain in such a way as to make π its invariant distribution. This method is essentially the *Metropolis-Hastings algorithm.* The other method simplifies the high dimensional problem by successively simulating from different smaller-dimensional components of $\mathbf{X} = (X^{(1)}, \dots X^{(d)})$. This idea is the essence of the *Gibbs sampler.* There is a wealth of different ways of applying these basic methods, and the two strategies can be combined in many different ways to suit different problems.

Within spatial statistics, π might define the distribution of a spatial process, or in other contexts, it might describe a Bayesian posterior distribution of parameters which govern an underlying spatial process. More complex distributions incorporating both these elements are also common. Since numbers of locations are typically very large, the dimensionality of

π is often extremely large, so that the need for MCMC methods in spatial statistics is particularly strong. Moreover, many spatial distributions exhibit strong and complex dependence structures, which can often severely inhibit MCMC performance.

In this chapter, we shall introduce MCMC methods in a general statistical context, although our illustrative example will have a simple spatial statistics structure that will be generalised considerably in later chapters. Our aim is not to give a comprehensive treatment of the subject, and interested readers wishing to learn more on this should read one of the currently available texts on the subject, see for example Gilks, Richardson & Spiegelhalter (1996) Tanner (1996), Gamerman (1997), Robert & Casella (1999), and Liu (2001). We will however provide sufficient details to give the reader an overview of the techniques, their immense generality and power, and some of the common pitfalls.

1.2 The Gibbs sampler

The most obvious way to break down a difficult high-dimensional simulation problem to a more tractable one, is to try to reduce it to a collection of more manageable smaller dimensional simulations. Many MCMC algorithms do this, but the simplest form of this strategy is an important special case called the *Gibbs sampler.*

Suppose that we wish to obtain a sample from the multivariate density $\pi(x^{(1)}, \ldots, x^{(d)})$. (We shall assume now and throughout this chapter, that distributions are continuous but everything extends immediately to the discrete distribution case.) We shall always use bracketed superscripts to index over components, so that a typical random variable distributed according to π might be written $\mathbf{X} = (X^{(1)}, \ldots X^{(d)})$, and the conditional distribution of the ith component given the rest according to π will simply be written as *the conditional* $X^{(i)} | (X^{(1)}, \ldots, X^{(i-1)}, X^{(i+1)}, \ldots, X^{(d)}$. The d conditional distributions of this type are termed the *full conditional distributions.*

The Gibbs sampler successively and repeatedly simulates from the conditional distributions of each component given the other components, see Algorithm 1 below. The realisations $\boldsymbol{X}_1, \boldsymbol{X}_2, \ldots$ obtained from iterating step 2 in Algorithm 1 are simulated from a Markov chain. Under mild regularity conditions (Roberts & Smith 1994), convergence of this chain to the stationary distribution $\pi(x^{(1)}, \ldots, x^{(d)})$ is guaranteed, so that for sufficiently large k, the values $(x_k^{(1)}, \ldots, x_k^{(d)})$, $\ldots, (x_n^{(1)}, \ldots, x_n^{(d)})$ can be regarded as realisations from this distribution.

Algorithm 1 *The Gibbs sampler.*

1. `Initialise with` $\boldsymbol{X}_0 = (X_0^{(1)}, \ldots, X_0^{(d)})$.

2. `Simulate` $X_1^{(1)}$ `from the conditional` $X^{(1)}|(X_0^{(2)}, \ldots, X_0^{(d)})$.

   `Simulate` $X_1^{(2)}$ `from the conditional` $X^{(2)}|(X_1^{(1)}, X_0^{(3)}, \ldots, X_0^{(d)})$.

   ...

   `Simulate` $X_1^{(d)}$ `from the conditional` $X^{(d)}|(X_1^{(1)}, \ldots, X_1^{(d-1)})$.

`Iterate this procedure.`

1.2.1 The Gibbs sampler and Bayesian statistics

Markov chain Monte Carlo was devised 50 years ago by Metropolis, Rosenbluth, Rosenbluth, Teller & Teller (1953) with a view for applications in statistical physics. It is a strange anomaly that its popularity within mainstream statistics is much more recent. The Gibbs sampler does appear in the statistics literature dating back to Suomela (1976); see the survey by Møller (1999). It became well-known in certain application areas, notably image analysis; see, for example, Geman & Geman (1984). However, its general popularity in statistics only developed as a result of Gelfand & Smith (1990). This paper only deals with applications of the Gibbs sampler (which even then was known to be a special case of MCMC) to Bayesian inference problems. It was even later that more general Metropolis-Hastings algorithms became widely adopted within statistical applications, thanks in particular to Besag & Green (1993).

Though many applications of Gibbs sampling exist outside the area of Bayesian statistics, Bayesian applications are undoubtably the most prolific. In this situation, the vector $\boldsymbol{X}$ represents a parameter vector, usually denoted by $\boldsymbol{\theta}$. Its appeal in the Bayesian context stems from the fact that its application is particularly natural in Bayesian problems which exhibit *conditional conjugacy.* Conditional conjugacy describes the common situation where the full conditional distributions to be sampled in Algorithm 1 above come from parameterised families of standard distributions, thus making the constituent steps in Algorithm 1 straightforward. The Gibbs sampler has proved particularly successful in its application to Bayesian analysis of classes of models which contain conditional independence structure, since this further simplifies the form of the full-conditional densities in Algorithm 1. Important examples include many classes of hierarchical models, see for example Smith & Roberts (1993), and missing data problems. See also BUGS (`http://www.mrc-bsu.cam.ac.uk/bugs/`) for software particularly suited for Gibbs sampling in hierarchical models.

1.2.2 Data augmentation

Data augmentation is a technique which can be thought of as a special case of the Gibbs sampler. It can be applied equally well to the following two situations.

1. One of the main problems that afflicts statistical computing is that some of the data could be missing. In theory the distribution of the observed data can be obtained by integrating out the missing data. However this is frequently difficult if not impossible to do.

2. In *latent models* such as mixture models and hidden Markov models, the likelihood can be expressed in terms of an expectation or an integral with respect to the latent structure, but often this expression cannot be calculated efficiently. However, given the latent structure, the conditional likelihood is frequently much simpler. In this situation it is useful to be able to treat the latent structure as unobserved *augmented data.*

Let $\boldsymbol{y}_{obs}$ denote the observed data, $\boldsymbol{y}_{mis}$ denote the missing data, and $\boldsymbol{\theta}$ be the unknown parameter of the likelihood with prior $\pi_0(\boldsymbol{\theta})$. Then the above examples demonstrate a situation in which $\pi(\boldsymbol{y}_{obs}, \boldsymbol{y}_{mis} \mid \boldsymbol{\theta})$ is available but $\pi(\boldsymbol{y}_{obs} \mid \boldsymbol{\theta})$ is not. As a result of this, we consider $\boldsymbol{y}_{mis}$ a further set of parameters so that

$$\pi(\boldsymbol{\theta}, \boldsymbol{y}_{mis}|\boldsymbol{y}_{obs}) \propto \pi(\boldsymbol{y}_{obs}, \boldsymbol{y}_{mis}|\boldsymbol{\theta})\pi_0(\boldsymbol{\theta}), \tag{1.1}$$

where π_0 is the prior density for θ. Data augmentation proceeds by carrying out Gibbs sampling to successively sample θ and $\boldsymbol{Y}_{mis}$ to produce a sample from this joint distribution. The marginal distribution of θ is therefore the posterior distribution of interest.

1.2.3 Gibbs sampling and convergence

Gibbs sampling pays a price for reducing the dimensionality of the simulations to be carried out. If the components, $X^{(1)}, \ldots X^{(d)}$ exhibit heavy dependence, the algorithm will converge slowly and give highly serially correlated output. A detailed description of the relationship between autocorrelation structure and convergence is beyond the scope of this chapter, though the reader is referred to Roberts & Sahu (1997) for further information. One practical point is that parameterisation choice is crucial to the convergence of the sampler, see for example Hills & Smith (1992) and Papaspiliopoulos, Roberts & Skold (2002).

1.3 The Metropolis–Hastings algorithm

The basic Metropolis–Hastings algorithm offers a flexible family of MCMC algorithms complementary to the Gibbs sampler. Later we will see how the ideas can be effectively combined, but here we shall introduce the Metropolis-Hastings method in its pure form.

Suppose that we wish to simulate from a (multivariate) distribution $\pi(x)$. We now let $q(x, y)$ be any arbitrary transition probability (that is $q(x, y)$ is the probability density of moving to y from x), but from which simulation is straightforward. Then the Metropolis–Hastings algorithm is:

Algorithm 2 *The Metropolis-Hastings algorithm.*

---

1. Given the nth value, $X_n = x$, generate a 'candidate value' y^* from density $q(x, \cdot)$.
2. Calculate $\alpha(x, y^*)$ where

$$\alpha(x,y) = \begin{cases} \min\{\frac{\pi(y)q(y,x)}{\pi(x)q(x,y)}, 1\} & \text{if } \pi(x)q(x,y) > 0 \\ 1 & \text{if } \pi(x)q(x,y) = 0 \end{cases}$$

3. With probability $\alpha(x, y^*)$ accept the candidate value and set $X_{n+1} = y^*$; otherwise reject and set $X_{n+1} = x$.
4. Repeat.

The Metropolis–Hastings algorithm has the major advantage over the Gibbs sampler that it is not necessary to know all the conditional distributions — we only need to simulate from q which we can choose arbitrarily, as long as q is capable of allowing us to reach all areas of positive probability under π (irreducibility). In fact, we only need to know π up to proportionality, since any constants of proportionality cancel in the numerator and denominator of the calculation of α. The price for the simplicity is that if q is poorly chosen, then either the proportion of rejections can be high, or the constructed Markov chain might move around the space too slowly. In both cases, this leads to the efficiency of the procedure for Monte Carlo estimation being very low. This problem is exacerbated by the fact that a particular choice of q might work well on one target density, but be extremely poor on another. Therefore it is impossible to prescribe application-independent strategies for choosing π.

In contrast to the Gibbs sampler (which uses explicitly the d–dimensional structure of the state space) the construction of Metropolis-Hastings given above needs no explicit structure of this type. Where the state space is (say) the d-dimensional Euclidean space, there is no requirement that each move update each of the d components simultaneously, and often smaller dimensional updates will prove to be computationally cheaper and almost as effective if the choice of components updated on each occasion is allowed to vary. In this way the Gibbs sampler can be seen to be a special case of Metropolis-Hastings in Section 1.3.2 and more general strategies will be discussed further in Section 1.5.2.

In the sequel, in our different examples of Metropolis-Hastings algorithms we consider for ease of exposition only the case where $\pi(x)q(x,y) > 0$ in step 2 of Algorithm 2.

1.3.1 *Examples of Metropolis–Hastings algorithms*

The Metropolis-Hastings framework is very general and flexible since it imposes no restriction whatsoever on the form of q except some mild regularity conditions that are required to ensure irreducibility and aperiodicity; see, for example, Roberts & Smith (1994). All currently used MCMC algorithms can be expressed in this form. Here we will discuss many examples of the most popular forms of Metropolis-Hastings algorithm, though the collection we describe is by no means exhaustive.

The independence sampler

The simplest possible choice for the proposal distribution chooses q to be *independent* of its first argument:

$$q(\boldsymbol{x}, \boldsymbol{y}) = q(\boldsymbol{y}) \;, \tag{1.2}$$

where we insist that the support of π is contained in that of q. Here we can write

$$\alpha(\boldsymbol{x}, \boldsymbol{y}) = \min\left\{\frac{w(\boldsymbol{y})}{w(\boldsymbol{x})}, 1\right\} \tag{1.3}$$

where

$$w(\boldsymbol{x}) = \frac{\pi(\boldsymbol{x})}{q(\boldsymbol{x})} \;. \tag{1.4}$$

It is clear that by taking q to be proportional to π, w is constant and the algorithm reduces, as we would expect, to IID sampling from π.

Symmetric random walk Metropolis algorithm

We define a random walk with transition density $q(\boldsymbol{x}, \boldsymbol{y}) = q_{rw}(\boldsymbol{y} - \boldsymbol{x})$ where we assume that q_{rw} is an even function ($q_{rw}(\boldsymbol{x}) = q_{rw}(-\boldsymbol{x})$ for all $\boldsymbol{x}$).

In this case the acceptance probability $\alpha(x,y)$ reduces to the simpler form

$$\alpha(\boldsymbol{x}, \boldsymbol{y}) = \min\left\{\frac{\pi(\boldsymbol{y})}{\pi(\boldsymbol{x})}, 1\right\} \;. \tag{1.5}$$

Thus the accept/reject mechanism can be interpreted as follows: accept all moves which increase π ("uphill moves"), but reject some moves which decrease π ("downhill moves"). The algorithm biases the random walk by moving towards modes of π more often than moving away from them.

Although a special case of the Metropolis-Hastings algorithm, random walk Metropolis still retains considerable flexibility in terms of choosing the random walk jump distribution $q_{rw}(\cdot)$. Furthermore, one appealing feature is that π only enters the algorithm in the accept/reject mechanism, making the algorithm extremely easy to implement. This has led to it becoming one of the most widely used MCMC methods.

Multiplicative random walk algorithms

An alternative choice to random walk Metropolis can be produced by considering a logarithmic random walk algorithm, where the proposed move is to a random multiple of the current state. This can be applied whenever the state space of the variable is contained in $\mathbf{R}^+$ From current state X_n we propose a move to $X_{n+1} = X_n e^Z$ where Z is drawn from a symmetric density q_{rw}. By a simple Jacobian transformation of π, the acceptance probability can be easily shown to be

$$\alpha(X_n, X_{n+1}) = \min\left\{1, \frac{\pi(X_{n+1})X_{n+1}}{\pi(X_n)X_n}\right\}.$$

Note that this algorithm is equivalent to that obtained by taking logs of a variable with distribution π and performing random-walk Metropolis on the resultant density. For situations where the support of π is not bounded away from zero, a more prudent transformation to use is $\log(a + x)$ for some suitable positive constant a. This is called a *modified multiplicative random walk* algorithm and avoids any potential problems caused by making arbitrarily small proposed moves for x near 0.

Langevin algorithms

Metropolis algorithms use no information from the target density in order to determine the location of sensible regions to move. Therefore they rely very heavily on the accept/reject mechanism to 'guide' the algorithm to the correct regions. This is often an adequate solution, but where further local information about the target density (such as its gradient) is available, this information can often be harnessed to produce more efficient methods.

The simplest such methods are the class of so-called Langevin methods (Besag 1994, Roberts & Tweedie 1996), which are motivated by continuous time diffusion processes which have stationary distribution π. The basic *Langevin algorithm* on $\mathbf{R}^d$ (sometimes called MALA (*Metropolis-adjusted Langevin algorithm*) or the *Langevin-Hastings algorithm*) uses a proposal distribution given by

$$q(\boldsymbol{x}, \boldsymbol{y}) = \frac{1}{(2\pi\sigma)^{d/2}} \exp\left\{\frac{-\|\boldsymbol{y} - \boldsymbol{x} - \sigma^2 \nabla \log \pi(\boldsymbol{x})/2\|^2}{2\sigma^2}\right\},$$

for some suitable scaling parameter σ. That is the proposed move from $\boldsymbol{x}$ is assumed to have a d-dimensional $N(\boldsymbol{x}+\sigma^2 \nabla \log \pi(\boldsymbol{x})/2, \sigma^2 I_d)$ distribution. Here ∇ just defines the usual gradient operator. Rather than be centred at the current state, the proposal centre is adjusted according to information about where the target density is likely to be larger.

The Langevin-Hastings algorithm has proved very successful in spatial applications, and will feature prominently in some of the applications of Sections 2.9 and 4.6.2. Theoretical results from Roberts & Tweedie (1998) suggest that Langevin methods in high dimensional applications ought to be substantially superior to the use of simpler alternatives such as the random walk Metropolis algorithm, and these results are borne out in the use of Langevin methods in Sections 2.9 and 4.6.2, when in many cases it is impossible to obtain adequate mixing for a random walk Metropolis algorithm.

As with the random walk Metropolis algorithm, a key problem is the identification of the a sensible value for σ^2. If σ^2 is too small, moves are invariably accepted, though the algorithm is inefficient since it moves around the space slowly. On the other hand, large values of σ^2 lead to most proposed moves being rejected. A practical problem is that the meaning 'large' and 'small' in this context depends crucially upon π. Theoretical results giving practical guidance about the scaling problem for the Langevin-Hastings algorithm is given in Roberts & Tweedie (1998), and scaling problems for more general algorithms are discussed in Roberts & Rosenthal (2001b).

Nothing comes for free of course, and the problem with implementing Langevin-Hastings comes from the need to compute derivatives of π. In the worst case, numerical approximations for the derivative of π may need to be computed. However often, especially in suitable exponential families, and where suitable conditional independence structure is present, an analytic expression for $\nabla \log \pi(\mathbf{x})$, will lead to this expression having a reasonably simple form. In this case, the computational cost of implementing Langevin-Hastings is usually relatively modest.

More sophisticated Langevin methods have even better theoretical properties than Langevin-Hastings (Stramer & Tweedie 1999) but often the implementation of these methods is far more time-consuming, involving the evaluation of complicated matrix exponentials at each iteration for their implementation.

1.3.2 The Gibbs sampler

In the form described above in Section 1.2, the Gibbs sampler is not strictly speaking a Metropolis-Hastings algorithm as described in Algorithm 2. However its d constituent update steps are all Hastings updates with acceptance rates identically equal to 1. This follows from setting $q_i(\mathbf{x}, \mathbf{y}) = \pi(y^{(i)} \mid x^{(1)}, \ldots x^{(i-1)}, x^{(i+1)}, \ldots x^{(d)})$ when

$$(x^{(1)}, \ldots x^{(i-1)}, x^{(i+1)}, \ldots x^{(d)}) = (y^{(1)}, \ldots y^{(i-1)}, y^{(i+1)}, \ldots y^{(d)})$$

and 0 otherwise. It is readily checked that with such a proposal, $\alpha \equiv 1$ in step 2 of Algorithm 2.

Auxiliary variable methods

In a number of different contexts, it turns out to be easier to simulate a Markov chain on a higher dimensional target distribution, from which π can be extracted as an appropriate marginal distribution. The concept of additional (or *auxiliary*) variables was introduced in the statistical physics literature by Edwards & Sokal (1988), though special cases were introduced earlier, see notably Swendsen & Wang (1987). The methods did not become widely known in statistics until Besag & Green (1993); see also Section 3.3.4. Auxiliary variables may have no interpretation related to the target density currently being investigated. They are usually introduced solely in order to improve MCMC mixing.

Suppose $\boldsymbol{X} \sim \pi$ and an auxiliary variable $\mathbf{U} \in \mathbf{R}^l$ is defined in terms of its conditional distribution given $\boldsymbol{X}$:

$$\mathbf{U}|\boldsymbol{X} = \boldsymbol{x} \sim h(\cdot|\boldsymbol{x}) \tag{1.6}$$

$(\boldsymbol{X}, \mathbf{U})$ is now an $\mathbf{R}^{d+l}$ random variable with joint density

$$\pi_E(\boldsymbol{x}, \boldsymbol{\beta}) \propto \pi(\boldsymbol{x}) h(\boldsymbol{\beta}|\boldsymbol{x}) \ . \tag{1.7}$$

Though apparently more complicated than π, π_E might be more suitable for exploration by MCMC. A popular application of auxiliary variables is the slice sampler; see Damien, Wakefield & Walker (1999) and Neal (2002). The slice sampler makes use of suitable factorisations of the density to introduce auxiliary variables to simplify the sampling problem. Suppose that π admits a factorisation

$$\pi(\boldsymbol{x}) = f_0(\boldsymbol{x}) \prod_{i=1}^{l} f_i(\boldsymbol{x}) \ . \tag{1.8}$$

Often very natural factorisations exist. For instance if π is the posterior density from a Bayesian analysis, f_0 could be the prior with f_i being the contribution to the likelihood from the ith data point, $1 \leq i \leq l$.

Now define $\mathbf{U}$ as l conditionally independent components with

$$U^{(i)}|\boldsymbol{X} \sim U(0, f_i(\boldsymbol{X})) \ . \tag{1.9}$$

Letting $S(\boldsymbol{\beta}) = \{\boldsymbol{x} : \; u^{(i)} \leq f_i(\boldsymbol{x}), 1 \leq i \leq l\}$, then it is easily seen that the joint density of $\boldsymbol{X}$ and $\mathbf{U}$ is given by

$$\pi_E(\boldsymbol{x}, \boldsymbol{\beta}) \propto f_0(\boldsymbol{x})\mathbf{1}[\boldsymbol{x} \in S(\boldsymbol{\beta})] \tag{1.10}$$

where $\mathbf{1}[\cdot]$ denotes the indicator function. Often we take f_0 to be constant, and in this case, π_E is the uniform density on $\cap_{i=1}^{l}\{\boldsymbol{x} : \; u^{(i)} \leq f_i(\boldsymbol{x})\}$.

The slice sampler carries out a Gibbs sampler on π_E. The conditional distribution of $\mathbf{U}|\mathbf{X}$ is specified by (1.9) and the requirement that the $U^{(i)}$s be conditionally independent. Therefore this part of the algorithm is straightforward to implement. The more challenging task is to sample from $\mathbf{X}|\mathbf{U}$ since it involves simulating from a density proportional to the truncated function $f_0(\boldsymbol{x})\mathbf{1}[\boldsymbol{x} \in S(\boldsymbol{\beta})]$. Even where f_0 is constant, this simulation can be difficult, since this involves simulating from the uniform distribution on a possibly high-dimensional irregularly shaped region, and this limits the applicability of the slice sampler. However Damien et al. (1999) and Neal (2002) offer a number of possible strategies for carrying out this step of the simulation.

Broadly speaking Neal (2002) concentrates on the case where $l = 1$, and describes methods for using rejection sampling to carry out the sampling of $\mathbf{X}|\mathbf{U}$. On the other hand, in Damien et al. (1999) we see that $l > 1$ is often used, with the choice of factorisation being dictated by the need to explicitly represent $S(\mathbf{U})$ as a constraint on possible possible state space values. This essentially requires us to be able to invert each of the functions f_i, $1 \leq i \leq l$.

1.4 MCMC Theory

To understand the theory of MCMC, it is imperative that we grasp certain basic concepts about Markov chains. In this section we will give the briefest of introductions to various important ideas in Markov chain theory. Going further into the theory would distract from the main themes of this book, but we will give certain key references. Just about all the concepts introduced here, and more, will appear in Roberts & Tweedie (2002).

1.4.1 Markov chains and MCMC

A *Markov chain* $\{X_n, \; n \geq 0\}$ is a discrete time random process with the following 'memoryless' property:

$$P[X_n \in A \mid X_{n-1} = x_{n-1}, X_{n-2} = x_{n-2}, \ldots, X_0 = x_0]$$

$$= P[X_n \in A \mid X_{n-1} = x_{n-1}].$$

Assume first that the state space, $\mathtt{X}$, is countable, and let $i, j \in \mathtt{X}$. Let $P_{ij} = P[X_n = j \mid X_{n-1} = i]$, and let P denote the matrix with elements $\{P_{ij}\}$. Then $(P^n)_{ij} = P[X_n = j \mid X_0 = i]$. A Markov chain is

irreducible if, for all i and j, there exists some n such that $P_{ij}^n > 0$;

aperiodic if the greatest common divider of $\{n : P_{ii}^n > 0\}$ is 1 for some i;

recurrent if P[Markov chain returns to $i \mid$ it started at i] $= 1$ for all i; and

positive recurrent if it is irreducible aperiodic and recurrent, and $\{\pi_j\}$ is a collection of probabilities (i.e. that sum to 1) such that

$$\lim_{n \to \infty} P_{ij}^n = \pi_j \tag{1.11}$$

and

$$\sum_i \pi_i P_{ij} = \pi_j \ . \tag{1.12}$$

Note that for the definition of aperiodicity, the choice of i is not important so long as the chain is irreducible.

These concepts are well-known from undergraduate probability material, however for MCMC theory we will need to consider more general state spaces than the discrete set up above. Assume now that we have a probability distribution $\pi(\cdot)$ on a state space $\mathtt{X}$. This probability distribution is typically the posterior distribution, but it is worth keeping in mind that $\mathtt{X}$ can be larger than the state space Θ of the parameters, for example it can include model indicators. This will be evident in Section 1.7.

It turns out that the general state space generalisations of irreducibility, aperiodicity and positive recurrence offer no surprises once the discrete concepts have been taken on board. The technicalities though do require some care. We refer the interested reader to Meyn & Tweedie (1993), Tierney (1994) or Roberts & Tweedie (2002) for details.

Denote by $P(x, \cdot) \equiv P(X_t \in \cdot | X_{t-1} = x)$ the *transition probability kernel* of the Markov chain. Now define the probability measure, νP by

$$\nu P(A) \equiv \int \nu(dx) P(x, A) \tag{1.13}$$

(for any set A) which is the distribution of the Markov chain at time t if the transition kernel is P and the distribution at time $t-1$ is ν for an arbitrary probability measure ν. A Markov chain has *stationary (or invariant) distribution* $\pi(\cdot)$ if

$$\pi P(\cdot) = \pi(\cdot) \tag{1.14}$$

which describes just the general state space version of (1.12). A sufficient, but not necessary, condition of a transition kernel to produce Markov chains with the required stationary distribution is *reversibility*. Reversibility implies that

$$\pi(dx)P(x,dy) = \pi(dy)P(y,dx) \tag{1.15}$$

which balances the probability of going from x to y with that from y to x. This equation is to be interpreted as an equality of probability measures: the joint distribution of the current state and the next is equivalent to that of the next and current. Thus the distribution proceeding forward in time is equivalent to that moving backwards in time, and this explains the use of the term reversibility. Reversibility implies that π is a stationary distribution since

$$\pi P(dy) = \int \pi(dx)P(x,dy) = \int \pi(dy)P(y,dx) = \pi(dy). \tag{1.16}$$

The relationship between Markov chains and MCMC methods emanate from the following crucial results. Assume that we can construct an aperiodic and irreducible Markov chain with state space $\mathtt{X}$, whose stationary distribution is $\pi(\cdot)$. Then, standard Markov chain theory tells us that for any initial seed x_0, the realisation of the chain $X_1, X_2, X_3, \ldots$ provides, through the ergodic theorem, a realisation of the stationary distribution since

$$X_t \to \pi \quad \text{in distribution as } t \to \infty \tag{1.17}$$

and an approximation of required expectations because

$$T^{-1}\sum_{t=1}^{T} g(X_t) \to \mathbf{E}_\pi[g(X)] \quad \text{almost surely, as } T \to \infty, \tag{1.18}$$

for any π-integrable function g.

For MCMC, the invariant probability distribution π is given – this is the distribution from which we wish to sample. Therefore it remains to demonstrate irreducibility and aperiodicity. They can usually be easily checked in individual situations, although there do occur examples where the use of MCMC is thwarted by reducibility, and others in which reducibility is difficult to acertain.

The following sections will investigate more closely convergence in distribution of the Markov chain and convergence of ergodic averages.

1.4.2 Convergence in distribution

Convergence to stationarity of Markov chains can be studied in a variety of different norms. Often convenient (though not always appropriate) is the *total variation distance* given for two probability measures ν_1 and ν_2, by

$$\|\nu_1 - \nu_2\| = \frac{1}{2}\int_{\mathtt{X}} |\nu_1(dx) - \nu_2(dx)| = \sup_{A\in\mathcal{B}} (\nu_1(A) - \nu_2(A)) \,. \tag{1.19}$$

When ν_1 and ν_2 admit densities with respect to Lebesgue measure f_1 and f_2 say, then, $\|\nu_1 - \nu_2\| = \frac{1}{2}\int_{\mathtt{X}} |f_1(x) - f_2(x)|dx$.

A minimal requirement for any sensible algorithm with transition probabilities P is that of *ergodicity*: that is for all $x \in \mathtt{X}$, as $n \to \infty$,

$$\|P^n(\boldsymbol{x}, \cdot) - \pi(\cdot)\| := r(\boldsymbol{x}, n) \downarrow 0 . \tag{1.20}$$

However, it is necessary to consider more refined conditions on $r(\cdot, \cdot)$ in order to compare different MCMC methods. Discussing rates of convergence is beyond the scope of this chapter, but the reader interested in knowing more about this and the way in which the rate of convergence of an algorithm affects observed empirical properties of MCMC output, is referred to Roberts (2002).

1.4.3 Central limit theorems

The MCMC algorithm's convergence properties are closely linked to those of its excursions away from central areas of $\mathtt{X}$ called small sets (which can be assumed to be compact regions for simple algorithms like the random walk Metropolis algorithm). This in turn is very closely linked to the existence of Central Limit Theorems (CLT) for the Markov chain, which are important for any Monte Carlo implementation. We say that a $\sqrt{T}$-CLT exists for a function f, if

$$T^{1/2}\left(\frac{\sum_{t=1}^{T} f(\mathbf{X}_t)}{T} - \mathbf{E}_\pi(f(\mathbf{X}))\right) \to N(0, \tau_f \mathrm{Var}_\pi(f(\mathbf{X}))) \tag{1.21}$$

in distribution as $T \to \infty$, for some $\tau_f \in (0, \infty)$, where τ_f denotes the *integrated auto-correlation time* for estimating the function f using the Markov chain P, and given by

$$\tau_f = 1 + 2\sum_{k=1}^{\infty} \rho_k^f . \tag{1.22}$$

ρ_k^f is the auto-correlation of $\{f(\mathbf{X}_i),\ i = 0, 1, 2, \ldots\}$ at lag k and when the chain is assumed to be started from stationarity. It is of course important to try to construct Markov chains which mix rapidly and have a small value for τ_f. However, even more fundamentally, it is crucial for dependable estimation that τ_f is finite.

Roughly speaking, geometric ergodicity (that is where $r(\mathbf{x}, n) \to 0$ geometrically quickly as a function of n) is enough to provide CLTs wherever we might expect, that is for all functions which possess finite second moments with respect to the target density π. This is demonstrated in Roberts & Rosenthal (1997). More subtle results are available when convergence is slower, see Jarner & Roberts (2002b) and Jarner & Roberts (2002a) for results in this direction.

1.5 Practical implementation

1.5.1 Sampling from full conditional densities

Gibbs sampling requires the ability to sample from all full conditional densities. If these densities are not of known form, this may be hard. When these densities are univariate, black-box algorithms that generate samples from any density can be used instead; for a rich collection of such algorithms see Devroye (1986). A black-box algorithm well-suited for Gibbs sampling was developed in Gilks & Wild (1992). It is suitable for *log-concave* densities (although it has been generalised for certain types of non log-concave densities) and it is based on the fact that any concave function can be bounded by piecewise linear upper and lower bounds (hulls), constructed by using tangents at, and chords between, evaluated points on the function over its domain. The algorithm has turned out to be very useful since in many popular models, such as the most common generalised linear and proportional hazard models, the resulting full conditionals are log-concave; see Dellaportas & Smith (1993).

1.5.2 More flexible MCMC algorithms

The basic MCMC methods introduced earlier can be combined in various ways to produce more complicated and flexible methods. In this section, we shall discuss some of these techniques.

If π is stationary for two Markov chain kernels, it is also stationary if we apply first one and then the other, or if we randomly choose which one to apply each time, etc. Since this also holds if we are considering an arbitrary number of kernels, this already gives us a flexible family of ways of producing kernels with stationary distribution π, which although not reversible in their own right, are composed of easy to understand reversible components. We have already seen an example of this in the form of the Gibbs sampler which is built by applying a sequence of Metropolis-Hastings steps. However, there are other important ways in which kernels are combined to create an even greater flexibility of techniques available to the MCMC user.

There are many ways to construct transition kernels which are modifications or extensions of more basic MCMC methods. Often these modified methods can accelerate the algorithm's convergence properties. However, such improved convergence sometimes comes at the cost of increased computing time, so it is not sufficient to carry out a purely theoretical comparison of Markov chain convergence in order to find the most efficient algorithm.

The most popular of these is the *Metropolis-within-Gibbs*[1] situation where not all of the full-conditional distributions can be easily simulated. Like the Gibbs sampler, the Metropolis-within-Gibbs method cycles through the components describing the current state. Suppose we can write $\boldsymbol{x} = (\boldsymbol{x}^{(1)}, \ldots, \boldsymbol{x}^{(d)})$. At the ith step, the Gibbs sampler updates $\boldsymbol{x}^{(i)}$ according to the conditional distribution $\pi(\boldsymbol{x}^{(i)}|\boldsymbol{x}^{(-i)})$ where $\boldsymbol{x}^{(-i)}$ denotes all components apart from the ith. The more general Metropolis-within-Gibbs method just updates the ith component according to a Metropolis-Hastings step which has $\pi(\boldsymbol{x}^{(i)}|\boldsymbol{x}^{(-i)})$ as its stationary distribution.

In some sense Metropolis-within-Gibbs can be though of as a special case of Metropolis-Hastings, though it is helpful to distinguish it from algorithms which make full-dimensional updates at each step. Its flexibility is key to its importance. Whilst sharing the advantages of the Gibbs sampler in breaking down a large-dimensional simulation into several smaller-dimensional ones, it is not constrained by the need to be able to simulate from the full conditional distributions. Thus it is common for an algorithm to use d update steps in which most involve pure Gibbs updates according to the suitable conditional distributions, whilst only certain problematic components are treated using this more general Metropolis-within-Gibbs procedure.

The Metropolis-within-Gibbs algorithms popularity stems from its ease of implementation and flexibility. Unlike the more generic symmetric random walk Metropolis algorithm, it allows the exploitation of tractability of certain conditional distributions while retaining much of the general applicability of the random walk algorithm.

Another popular strategy is *blocking*: instead of sampling each component of $\mathbf{X}$ separately from its full conditional density, sample simultaneously from a subvector of $\mathbf{X}$, say $\mathbf{X}^{(s)}$. This provides an algorithm which is often superior to original Gibbs sampling (Roberts & Sahu 1997), though block updating can prove to be computationally too expensive in certain cases; though see Rue (2001) for computationally efficient ways of performing block updates where significant conditional independence is present.

Blocking can be combined with the Metropolis-within-Gibbs algorithm where perhaps the blocks are each updated according to multi-dimensional Metropolis-Hastings steps. One useful practical strategy is as follows. Approximate the covariance matrix of the target distribution by some means, and then tune the random walk Metropolis proposal density (or independence sampler proposal density) to have covariance proportional to the estimated target covariance. Estimating the target covariance could be carried

[1]The name Metropolis-within Gibbs derives from early applications where most components could be updated using Gibbs sampling, but a small number of complex full-conditionals existed for which Metropolis routines were loted in to replace the difficult Gibbs update step. Like the Gibbs sampler itself, its name is somewhat of a misnoma since it can easily be described as a special case of Metropolis-Hastings.

out on the basis of a pilot MCMC run, or in some cases, analytic tractability of (for instance) the Hessian of the target density at its mode can be used to estimate the inverse covariance.

Another useful technique which is usually effective when it can be practically applied, is to marginalise out particular components, and then to construct an algorithm on the resultant lower-dimensional space. Where Gibbs sampling is feasible in this lower-dimensional space, this algorithm is often termed the *collapsed Gibbs sampler*. Thus, instead of simulating, for example, $X^{(1)}$ from the conditional density $X^{(1)}|(X^{(2)}, \ldots, X^{(d)})$, we may instead sample from $X^{(1)}|\mathbf{X}^{(s)}$ (where $\mathbf{X}^{(s)}$ is a subset of $(X^{(2)}, \ldots, X^{(d)})$) produced by integrating out the remaining parameters. Finally, a rather important issue that should be carefully dealt with is *reparameterisation*. It is well known that strong posterior dependence among the co-ordinates of $\mathbf{X}$ produce slow mixing MCMC algorithms; see, for example, Roberts & Sahu (1997), and Section 4.6.3. This problem may be very severe in data augmentation problems involving missing or latent data, where the statistical information about the unknown parameters contained in the latent (imputed) data is considerably greater than that actually contained in the observed data. This problem is very closely connected to similar difficulties encountered in the implementation of the EM algorithm and its variants (Meng & van Dyk 1997).

One popular example that demonstrates very well this problem is the hierarchically *centred* algorithm in Bayesian linear models. Specifically, suppose that we have the simple structure

$$Y_i \sim (\theta_i, \sigma_e^2), \quad i = 1, 2 \ldots n \ ,$$

with a normal prior on the random effects,

$$\theta_i \sim N(\mu, \sigma_\alpha^2), \quad i = 1, 2 \ldots n$$

and an improper prior for μ. In this case with parameters $\{\theta_i, \ i = 1, \ldots n\}$ and μ, the Gibbs sampler which deterministically updates each parameter in turn can be shown (Roberts & Sahu 1997) to converge at a geometric convergence rate given by

$$\kappa = \frac{\sigma_e^2}{\sigma_e^2 + \sigma_\alpha^2},$$

so that $r(\mathbf{x}, n) \leq V(\mathbf{x})\kappa^n$ for some positive function V. In this context the latent layer consists of the random effects given by θ.

In summary, the centred parameterisation algorithm converges very slowly when measurement error (σ_e^2) is large in comparison to the latent layer variance (σ_α^2). However, there is a simple reparameterisation which permits good convergence whenever the measurement error is comparatively large. This is the *non-centred* parameterisation:

$$Y_i \sim (\mu + \theta_i, \sigma_e^2), \quad i = 1, 2 \ldots n \ ,$$

where each of the θ_i's are given independent priors with zero mean. The algorithm which updates each of the parameters in this parameterisation in turn according to a Gibbs sampler can be shown to have rate of convergence $1-\kappa$; again see Roberts & Sahu (1997). Similar non-centred algorithms can be applied in Poisson random effect models and elsewhere.

1.5.3 Implementation and output analysis

A real danger of MCMC is that it can be applied to virtually any model, whether or not the model is statistically sensible. Algorithms can produce very reasonable results, even when the number of parameters in the model exceeds by far the size of the data set. A general problem is that proper prior densities can 'hide' problems of model identifiability since the resulting posterior densities are proper. A way to avoid this is to always perform a sensitivity analysis to prior assumptions which includes remodelling with non-informative priors.

An important problem concerning the implementation of MCMC is to gauge when *convergence* has been 'achieved'; that is, to assess at what point the distribution of the chain is sufficiently close to its asymptotic distribution for all practical purposes. The number of iterations, B, needed to satisfy this is called the *burn-in* period or the *initial transient phase.*

It is usually sensible that sampled values within this (estimated) initial transient phase, B, are discarded for the purposes of Monte Carlo estimation, so the bias caused by the effect of starting values is reduced. A related problem tries to assess how many iterations are need for $(T-B)^{-1}\sum_{t=B+1}^{T} g(X_t)$ to estimate $\mathbf{E}_\pi[g(X)]$ sufficiently accurately for any function of interest g.

Particularly in high dimensions, assessment of convergence of MCMC by empirical observation is extremely difficult, and it is rarely possible to be absolutely certain that our algorithm has converged. Nevertheless, a number of useful *convergence diagnostic* procedures have been proposed, most of which involve the monitoring of a suitable function of the Markov chain, the monitored statistic also being termed a convergence diagnostic. It makes sense to use multiple (and diverse) convergence diagnostic procedures, so that the chance of discovering convergence problems is maximised. We shall not refer to these diagnostics here; for extensive reviews and comparisons see Cowles & Carlin (1996), Brooks & Roberts (1998) and Mengersen, Robert & Guihenneuc-Jouyaux (1999). Of course it must be emphasised that convergence diagnostics cannot guarantee convergence in any way. However, used carefully, they are an important practical tool.

Here though we content ourselves with a short discussion of sensible procedures for minimising the chance of failing to observe that the Markov chain has not converged. Apart from the formal convergence tests reviewed for example in Cowles & Carlin (1996), there exist a series of important informal convergence checks that can assist in convergence assessment. Visual inspection of trace plots from different starting values is an important initial investigation since it might reveal possible anomalies, for instance involving multimodality. Every effort should be made to ensure that the Markov chain is able to explore the entire state space. As a result of this, Gelman & Rubin (1992) strongly advocate the use of starting points for the replicated simulations which are over-dispersed (perhaps an estimate of 3 to 4 standard deviations away from the mean) with respect to the stationary distribution. When chains of length T from two starting values have visited substantially different regions of the state space, the burn-in period for at least one of the chains will be greater than T. The Gelman and Rubin procedure attempts to maximise the chance of detecting convergence problems of this type by using a collection of starting values which are as diverse as possible. Note that even the problem of obtaining over-dispersed starting values can be difficult when little is known about the target density.

Auto-correlation and *cross-correlation* plots of particular components also provide good indications of the chain's mixing behaviour. For functions f and g which have finite variance with respect to π, we set r_k^f to be the sample correlation of $\{(f(X_t), f(X_{t+k})),\ 1 \le t \le T-k\}$, and $r_k^{f,g}$ to be the sample correlation of $\{(f(X_t), g(X_{t+k})),\ 1 \le t \le T-k\}$. The (empirical) auto-correlation function of $f(X)$ plots r_k^f against k for $k = 0, 1, 2, \ldots$, while the (empirical) cross-correlation function of $f(X)$ and $g(X)$ plots $r_k^{f,g}$ against k for $k = 0, 1, 2, \ldots$. Both functions should normally be computed after elimination of the burn-in period.

These plots estimate the theoretical auto-correlation and cross-correlation plots, $\{\rho_k^f, k = 0, 1, 2, \ldots\}$ and $\{\rho_k^{f,g}, k = 0, 1, 2, \ldots\}$ respectively, and are consistent in the sense that $\lim_{T\to\infty} r_k^f = \rho_k^f$ almost surely and $\lim_{T\to\infty} r_k^{f,g} = \rho_k^{f,g}$ almost surely by applications of (1.18). $\rho_0^f = 1$ and for an ergodic Markov chain $\lim_{k\to\infty} \rho_k^f = 0$. For a rapidly mixing chain, the function ρ^f decays rapidly to 0 for all functions f, while for a slowly mixing chain the decay to zero is typically much slower, at least for some functions. A similar situation applies to the cross-correlation function $\rho^{f,g}$ although in this case $\rho_0^{f,g}$ can take any value in $[-1, 1]$. Therefore an indication as to the speed of mixing can be obtained by inspecting the auto-correlation and cross-correlation plots.

An important choice that arises in many applications of Metropolis-Hastings algorithms is the choice of the spread of the transition probability density $q(y, x)$ in Algorithm 2. For instance, if a proposal variance is too small for a particular target distribution, Markov chain mixing will be slow since only small jumps will be attempted. However, for an excessively large proposal variance, algorithms tend to be 'over ambitious', rejecting a large proportion of attempted transitions.

Not much is known about the problem of choosing proposal variances in general, but in the symmetric random walk case, some guidelines can be given. A sensible way to specify a spread is to tune the acceptance rate:

$$a = \int\int \pi(x)\alpha(x,y)q(x,y)dydx, \tag{1.23}$$

that is, a described the proportion of accepted moves for the algorithm. There are theoretical results in Gelman, Roberts & Gilks (1996) and Roberts, Gelman & Gilks (1997), which support the use of acceptance rates between 0.23 and 0.45, with the lower rates being preferred for higher dimensions. This theory is based largely on asymptotic arguments for large dimensional problems, so its application should not be carried out too rigidly. A sensible rule of thumb is to tune the acceptance rate to be between 0.20 and 0.50.

As for any *Monte Carlo estimator*, when an expectation $\mathbf{E}_\pi(f(\mathbf{X}))$ is estimated by (1.18) to give an estimator $\hat{\mathbf{E}}_\pi(f(\mathbf{X}))$, it is useful to also estimate its squared standard error, $\mathrm{Var}_\pi(\hat{\mathbf{E}}_\pi(f(\mathbf{X})))$. However, since the sample is not independent, this is not a simple issue even when a CLT for the Markov chain and the function of interest is known to exist. Geyer (1992) suggested the use of certain methods which have their origins in the time-series literature to calculate the standard error of the estimates. The simplest of these methods is the method of batch means which proceeds as follows. Divide the time series of length T into k batches $B_1, B_2, \ldots, B_k$ of equal size m and calculate the batch mean estimates by $\bar{B}_j = \hat{\mathbf{E}}_\pi^{(j)}(f(\mathbf{X})) = m^{-1}\sum_{i\in B_j} f(\mathbf{X}_i)$. Then, an estimate of $\mathrm{Var}_\pi(\hat{\mathbf{E}}_\pi(f(\mathbf{X})))$ is given by $k^{-1}(k-1)^{-1}\sum_{j=1}^k(\bar{B}_j - \bar{B})^2$, with $\bar{B} = k^{-1}\sum_{j=1}^k B_j$. Of course, the batch means are treated as independent and therefore the length m of each batch should be as large as possible, keeping in mind that slow mixing chains require larger batch sizes. One way to test the independence between batch means is to require the correlation between $\bar{B}_j$ and $\bar{B}_{j+1}$, $j = 1, 2, \ldots, k-1$, to be between ± 0.1.

An alternative method to estimate the standard error is to use the notion of *effective sample size* for estimating the expectation of a function $f(X)$ under π; see, for example, the discussion in Kass, Carlin, Gelman & Neal (1998). It is defined as

$$ESS = \frac{T}{\tau_f} \tag{1.24}$$

where T is the number of iterations and τ_f is the integrated auto-correlation time defined in (1.22).

Although r_k is a consistent estimator of ρ_k, $1 + 2\sum_{k=1}^{T-1} r_k$ for all k, is certainly not consistent for estimating τ_f, so that care has to be taken in estimating τ_f. A simple practical solution (which can be applied when the simulated Markov chain is reversible) was proposed in Geyer (1992), and involves estimating τ_f by $1 + 2\sum_{k=1}^{C} r_k^f$ where the truncation point C is chosen to be the largest k that $r_{2k+1}^f + r_{2k}^f > 0$ holds.

1.5.4 *Uses of MCMC in classical statistics and beyond*

MCMC has certainly had a fundamental effect on statistics in the last 15 years, influencing not only the way quantities for inference are computed, but also the type of models the statistician tries to fit. This is because the flexibility of the technique expands immeasurably the classes of statistical models for which computation is considered possible. This effect is most pronounced in Bayesian statistics. However MCMC also has many important applications in classical statistics. We shall briefly mention four examples.

1. Firstly note that given a flat prior, the likelihood and the posterior densities are equal. Therefore it is feasible to estimate the maximum of the likelihood as the mode of an estimated posterior density, assuming of course that the latter is integrable with respect to the parameters, which is the case if the likelihood is well defined. Thus, it is required to obtain an estimate of the density from a sample of dependent points from that density. This is usually done in terms of a *kernel density estimate*. For example, given a sample $x_i, 1 \leq i \leq n$ we could estimate the density as

$$\frac{1}{n}\sum_{i=1}^{n}\frac{1}{(2\pi\sigma^2)^{1/2}}\exp\{-(x-x_i)^2/(2\sigma^2)\} \tag{1.25}$$

   Maximisation of such a kernel density estimate usually needs to be performed using numerical or graphical methods, but for a one-dimensional problem, this does not cause any severe computational complications. The 'smoothness' parameter σ^2 needs to be chosen. As a general rule, the larger n, the smaller we would normally like to choose σ^2. It is possible to apply the theory of optimal scale choice for kernel density estimates for dependent data to this problem. It turns out that for Metropolis-Hastings algorithms, optimal choices of σ are larger than those for independent data, see Roberts & Skold (2001), though a detailed discussion of this area is beyond the scope of this chapter.

2. A rather different though not uncommon situation is when the likelihood given the unknown parameter is known only up to a normalisation constant. The idea is due to Geyer & Thompson (1992). Let

$$L(\theta|x) = \frac{h_\theta(x)}{c(\theta)} \tag{1.26}$$

   where $c(\theta)$ is unknown, but its value is crucial in determining maximum likelihood values for θ. In this case, it is necessary to estimate c. Now suppose θ_0 is a 'first guess' of the MLE. We can write

$$\frac{L(\theta|x)}{L(\theta_0|x)} = \frac{h_\theta(x)}{h_{\theta_0}(x)}\frac{c(\theta_0)}{c(\theta)}, \tag{1.27}$$

and can estimate $c(\theta)/c(\theta_0)$ for an arbitrary θ from a sample $(x_1, \ldots x_n)$ from h_{θ_0} using the estimator

$$\frac{1}{n}\sum_{i=1}^{n}\frac{h_\theta(x_i)}{h_{\theta_0}(x_i)}\,, \tag{1.28}$$

which is an unbiased estimator of $c(\theta)/c(\theta_0)$. From this simulation therefore, we can estimate $L(\theta|x)/L(\theta_0|x)$ from (1.28), and improve our estimate of the MLE. This procedure is usually iterated a number of times until the estimated terms stabilise. See also Sections 3.4.3 and 4.7.4.

3. *Simulated annealing* is a stochastic algorithm for global maximisation of functions. It is very closely related to MCMC. The only difference is that the target density changes as the algorithm proceeds. First notice that if we want to maximise a non-negative function $h(y), y \in A$, the maximising value also maximises $h(y)^{1/T}$ for arbitrary $T > 0$ known as the temperature. However for small T, assuming that $h^{1/T}$ is an integrable function, the density proportional to $h(y)^{1/T}$ concentrates most of its probability mass in the vicinity of the modes of the function. The idea behind simulated annealing is, at each iteration n, to carry out MCMC on a density proportional to h^{1/T_n}, where $\{T_n\}$ is a sequence of temperatures that converges to zero. The algorithm converges to the maximum value of h under suitable regularity conditions. It is important that T_n does not converge to zero too rapidly, since it turns out that this could then lead to the algorithm getting stuck in a minor mode of the function h. See Section 3.4.2.

4. MCMC is also used just to simulate from some stochastic models. Good examples include the point process models introduced in Chapter 4.

1.5.5 Software

A popular program to perform MCMC in a variety of models is BUGS (Bayesian inference Using Gibbs Sampling). It has been developed at the MRC biostatistics unit at the University of Cambridge and its recent version is called WinBUGS. Convergence diagnostics S-plus functions that may, or may not, accompany BUGS are CODA and BOA. For more information see at the web pages `http://www.mrc-bsu.cam.ac.uk/bugs/` and `http://www.public-health.uiowa.edu/boa/`.

1.6 An illustrative example

In this section we analyse a simple spatial problem to illustrate various MCMC techniques. Consider a 5×5 grid with data points $\boldsymbol{y} = (y_{11}, y_{12}, \ldots, y_{5,5})^T$ and assume that $\boldsymbol{y} \sim MVN(\mathbf{0}, \beta V)$ where MVN denotes the multivariate normal distribution, $\mathbf{0}$ is a 25×1 vector of zeros, $\beta > 0$ is an unknown scalar and V is the correlation matrix. V is parameterised by a parameter $\alpha \geq 0$ so that the correlation between two grid points $i = (i_1, i_2)$ and $j = (j_1, j_2)$ is $\exp\left(-\alpha^{-1}\|i - j\|\right)$ with $\|\cdot\|$ denoting the Euclidean distance on the spatial grid. The data $\boldsymbol{y}$ that fill the grid were simulated by using $\alpha = 5$ and $\beta = 1$.

For our Bayesian analysis, we assume the following independent priors: $\alpha \sim Ga\left(10^{-3}, 10^{-3}\right)$ and $\beta \sim IG\left(10^{-3}, 10^{-3}\right)$, where Ga and IG denote the Gamma and inverse Gamma distributions respectively with the notational convention that α and β are a priori identically distributed with mean 1 and variance 10^3.

The posterior density of (α, β) is given by

$$\begin{aligned}\pi(\alpha, \beta \mid \boldsymbol{y}) \quad &\propto \quad \beta^{-\frac{25}{2}} |V|^{-1/2} \exp\left\{-\frac{1}{2}\beta^{-1}\boldsymbol{y}^T V^{-1}\boldsymbol{y}\right\} \times \\ &\qquad \alpha^{-0.999} \exp\{-0.001\alpha\}\, \beta^{-1.001} \exp\left\{-0.001\beta^{-1}\right\}.\end{aligned}$$

The parameter β satisfies the conditional conjugacy property in this situation, and its distribution conditional on α can be written as

$$\pi(\beta|\alpha, \boldsymbol{y}) \propto IG\left(\frac{25}{2} + 10^{-3}, \frac{1}{2}\boldsymbol{y}^T V^{-1}\boldsymbol{y} + 10^{-3}\right). \tag{1.29}$$

Therefore sampling from this conditional distribution is straightforward. The conditional density of α given β is given by

$$\pi(\alpha \mid \beta, \boldsymbol{y}) \propto |V|^{-1/2} \exp\left\{-\frac{1}{2}\beta^{-1}\boldsymbol{y}^T V^{-1}\boldsymbol{y}\right\} \alpha^{-0.999} \exp\{-0.001\alpha\} \tag{1.30}$$

which is not of standard form, and so sampling from it is more problematic.

1.6.1 Metropolis-within Gibbs

The algorithm naturally suggested by this discussion is a Metropolis-within-Gibbs algorithm which samples β from its full conditional density given in (1.29) while a Metropolis step is used to update the α parameter with a random walk Metropolis step with stationary density given by (1.30). A normal proposal density was used, scaled appropriately in order to make the resulting acceptance rate about 0.4 as discussed above.

The resulting Metropolis-within-Gibbs algorithm produces rather disappointing results with highly serially correlated output as shown in the auto-correlation plot of Figure 1.1. The auto-correlation decays very slowly so

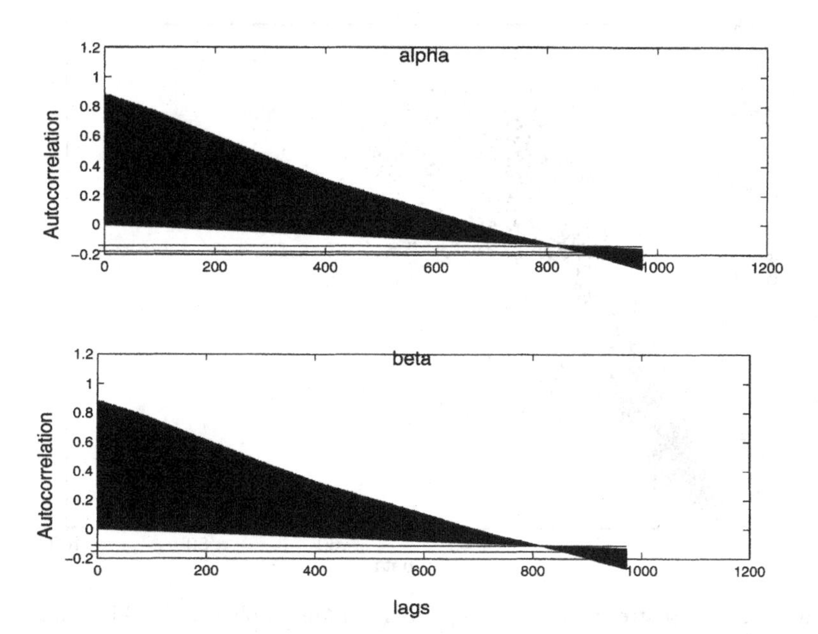

FIGURE 1.1. Autocorrelation functions of α and β. MCMC output is produced by sampling in turn from $\pi(\alpha \mid \beta, \boldsymbol{y})$ and $\pi(\beta \mid \alpha, \boldsymbol{y})$.

the mixing of this algorithm is poor, and estimation based on output from this Markov chain would be inefficient at best, not to mention potentially unreliable.

A glance at Figure 1.2 shows why the convergence of the algorithm is so poor. The parameters are highly correlated, so that a Gibbs sampling based strategy using this parameterisation will be extremely slowly mixing. A reparameterisation of α, β in order to reduce posterior correlations is a possible solution. However in this context, there is no natural class of reparameterisations from which to choose, leading us to suggest alternative remedies.

1.6.2 A collapsed algorithm

In this context, a much more promising avenue is to integrate out β leaving the marginal posterior distribution of α (marginalisation of α). Then a Markov chain on this one-dimensional problem can be constructed. As pointed out in Section 1.5.2 the resulting algorithm could be expected to converge more rapidly than the 2-dimensional sampler described above. Given a sample of values from the marginal distribution of α, $\alpha_1, \dots, \alpha_T$ say, we can simulate a corresponding sample of β values, $\beta_1 \dots, \beta_T$ respectively, with β_i drawn from $\pi(\beta|\alpha_i, \mathbf{y})$. For assessing convergence of the (α_i, β_i) sequence, it is sufficient to consider just the α_i's, since the β_i's are

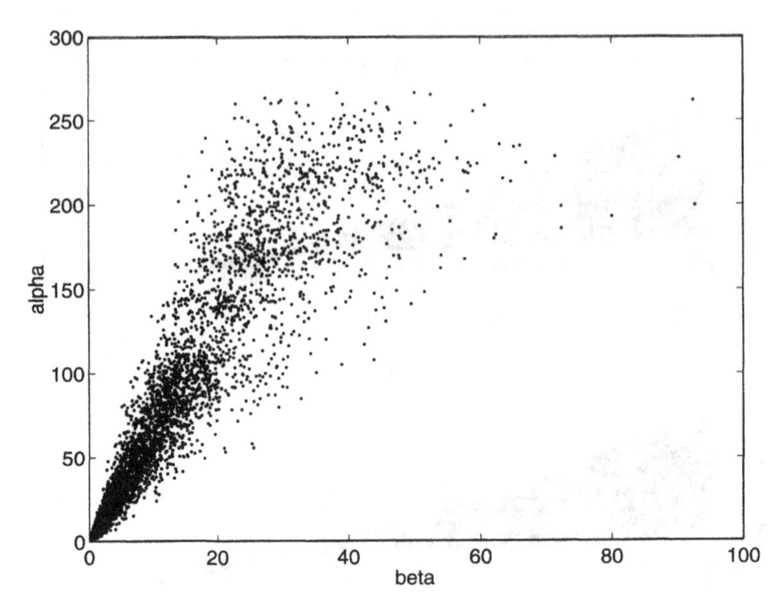

FIGURE 1.2. Scatter plot of α and β from the Metropolis-within-Gibbs output

drawn automatically from the correct distribution once α has converged to its marginal posterior distribution; see Roberts & Rosenthal (2001a).

A simple choice for the algorithm to explore $\pi\left(\alpha|\boldsymbol{y}\right)$ is the symmetric random walk Metropolis algorithm with normal proposal density. In our simulations we tuned the proposal variance to have acceptance rate of roughly around 0.33. The auto-correlation function of the resulting algorithm is depicted in the lower part of Figure 1.3. Once again, the auto-correlation function decays quite slowly, and in fact standard convergence tests (such as those available in CODA software) indicate that convergence has not been achieved after 600000 iterations, suggesting that other Metropolis-Hastings algorithms ought to be attempted.

1.6.3 The independence sampler

One obvious alternative is the independence sampler on the marginal posterior density of α. For the independence sampler, best results are obtained when the proposal density resembles the target density as closely as possible. However since mimicking the target distribution exactly is impossible, it is important to try to err on the side of having a proposal density which is over-dispersed with respect to the target density. This is usually feasible in a one-dimensional example such as this. In this example a mixture density of the form $0.45G\left(2.1, 1.5^{-1}\right) + 0.25G\left(1, 10^{-1}\right) + 0.3G\left(1, 70^{-1}\right)$, with $G(a,b)$ denoting the appropriate Gamma p.d.f., resembles $\pi(\alpha|\boldsymbol{y})$ quite closely but

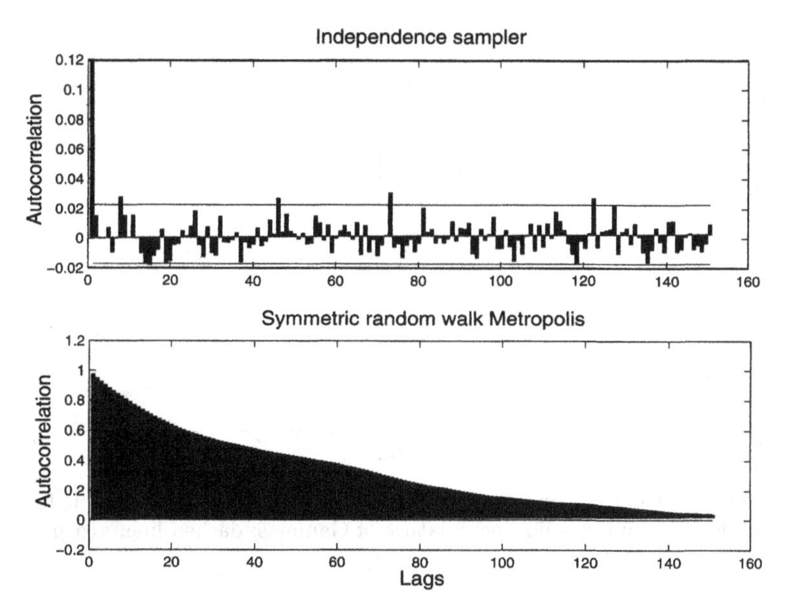

FIGURE 1.3. Autocorrelation functions of α for the independence sampler and the symmetric random walk Metropolis

possesses a right hand tail which is heavier than π as desired; see Figure 1.4. The resulting independence sampler produces the auto-correlation function shown in Figure 1.3 and acceptance rate 0.76 — very high for this algorithm.

Figure 1.5 depicts sections of trace plots of $\log \alpha$ for both the independence sampler and the symmetric random walk Metropolis algorithms of Section 1.6.2. It is clear that the random walk Metropolis algorithm exhibits excursions into the right tail of $\pi(\alpha \mid \boldsymbol{y})$ which are unpredictable and irregular. This is a common phenomenon for the random walk Metropolis algorithm in densities with heavy tails, and leads to further problems of unreliable Monte Carlo estimation; see, for example, Roberts (2002). The worst effects of this can be alleviated by the use of heavy tailed random walk proposals (Jarner & Roberts 2002a).

On the other hand, the independence sampler seems to have short but frequent excursions into the tails, indicating that inference from this sampler ought to be more reliable.

Another way to investigate the behaviour of these algorithms is to split the MCMC output into batches of fixed size, perhaps each of length 100 iterations. For each batch we compute a batch mean, B_i say and a batch acceptance rate a_i say. Finally we produce a scatter plot of (B_i, a_i) values; see Figure 1.6.

Overall, the acceptance rate of the independence sampler should be as

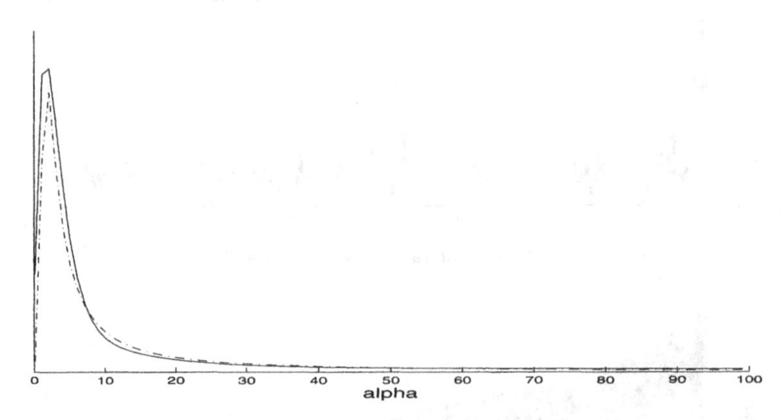

FIGURE 1.4. Comparison of $\pi(\alpha|\boldsymbol{y})$ and the proposal density chosen for the independence sampler; solid line: mixture of Gammas; dashed line: $\pi(\alpha|\boldsymbol{y})$.

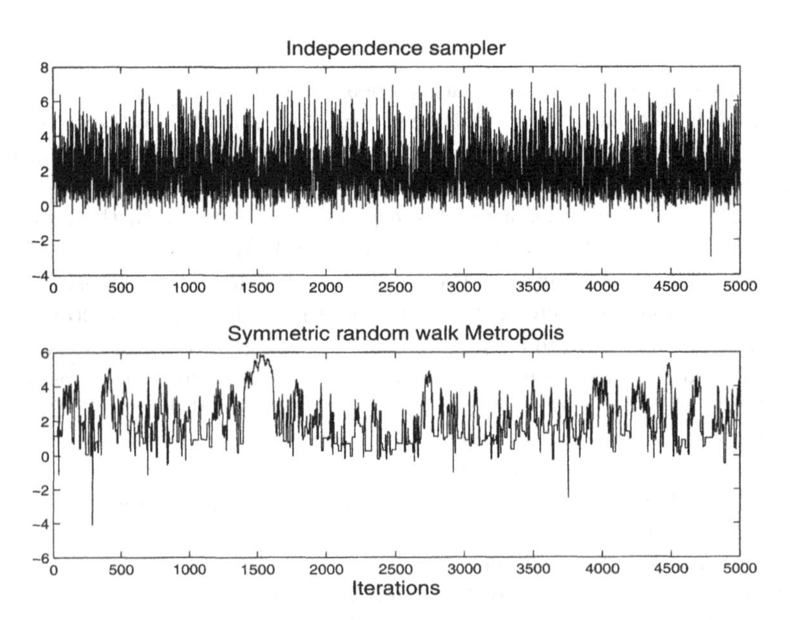

FIGURE 1.5. Trace plots of $\log \alpha$ for the independence sampler and the random walk Metropolis algorithm.

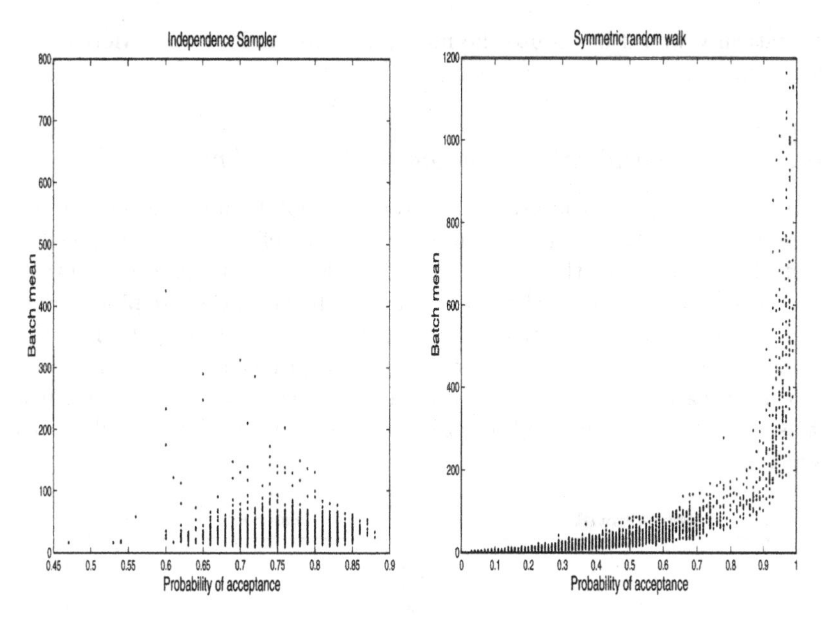

FIGURE 1.6. Acceptance rates against batch means for batch sizes 100.

high as possible, while that of the random walk algorithm should normally be between 0.2 and 0.5. However the plots in Figure 1.6 look more deeply into the way in which the acceptance rate varies across the state space. This can be useful in diagnosing convergence problems not obviously evident from other empirical investigations, and moreover can often be used to suggest better samplers, for instance samplers which behave differently in different parts of the space.

Figure 1.6 shows that the independence sampler has low acceptance rates in the left tail of the density, indicating that for values of α close to zero the proposal and the target densities do not match well and acceptance rates are very low leading to a 'sticky patch' in the state space. Since π is unbounded for α near 0, this is to be expected. Thus although the sampler's overall performance seems good for this particular MCMC simulation, an independent simulation might well get stuck near 0 for very large numbers of iterations, thus making Monte Carlo estimation of any quantities of interest very inaccurate.

On the other hand, the random walk Metropolis is problematic at the right tail of $\pi(\alpha \mid \boldsymbol{y})$ since the high acceptance probabilities indicate that the algorithm behaves very much like a random walk (with no rejections at all). This is caused by the fact that this heavy-tailed target density is extremely flat in this tail region. Random walks are known to have very heavy-tailed and hence unstable excursions, and this is consistent with that observed in Figure 1.5. A natural remedy here would be to propose larger moves in the right hand tail region, and a natural way of ensuring this automatically would be to use the multiplicative random walk Metropolis algorithm.

1.6.4 *The multiplicative random walk algorithm*

For positive random variables, it is often prudent to use the multiplicative random walk Metropolis to guard against the effects of a heavy-tailed target distribution. For this example, after 300000 iterations, the resultant acceptance rate against batch means graph is shown in the left plot of Figure 1.7. Now, the probability of acceptance does not converge to 1 in the right tail. However there are some very high acceptance rates for very small batch mean values indicating perhaps that the algorithm might get stuck in this region, proposing (and almost always accepting) very small moves at each iteration.

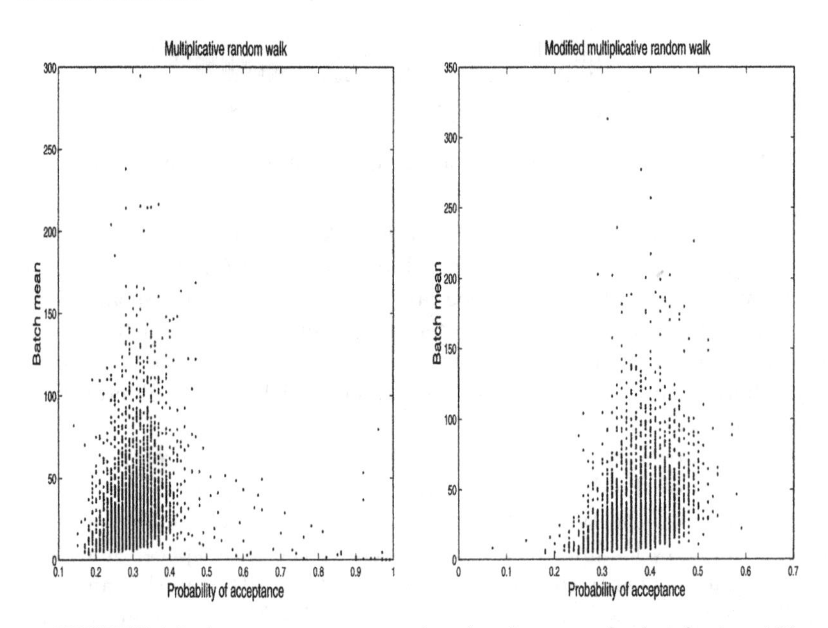

FIGURE 1.7. Acceptance rates against batch means for batch sizes 100.

Therefore it seems appropriate to also attempt the modified multiplicative random walk Metropolis, which operates a symmetric random walk on $\log(1 + \alpha)$. This method does not involve constraining proposal variances to go to zero for small values of α. The resulting acceptance rates vs batch means graph is shown in the right hand plot of Figure 1.7 and indicates a clear improvement over the standard multiplicative random walk Metropolis since the acceptance probabilities remain mainly in the desired range 0.2-0.5. To complete the picture, the corresponding (log) trace plots and auto-correlation functions for the multiplicative random walk algorithms are depicted in Figures 1.8 and 1.9 respectively.

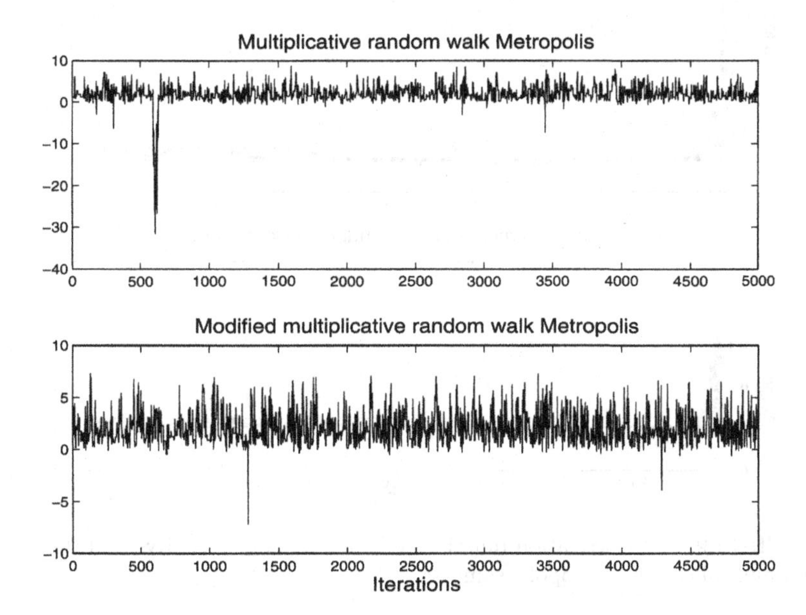

FIGURE 1.8. Trace plots of $\log \alpha$ for the multiplicative and the modified multiplicative Metropolis sampler.

Finally, note that in this illustrative example the target density that we need to sample from is univariate, and thus easy to investigate. In most cases, and in particular in Metropolis-within-Gibbs situations, such detailed descriptive analysis is impossible and what is left to the MCMC user is the ability to repeatedly run different algorithms with different starting points and investigate carefully diagnostic plots. However, the general principles discussed in this section regarding the use of independence samplers, random walk samplers, heavy tailed distributions, etc. carry over to more complex problems.

1.7 Appendix: Model determination using MCMC

In this appendix, we shall give a brief tour of Bayesian model selection methodology which is an important current research topic. Though not central to spatial statistics, many of the techniques described here are also of use in many spatial problems, and are used from time to time throughout the book. Our presentation will describe the methods in a general Bayesian setting.

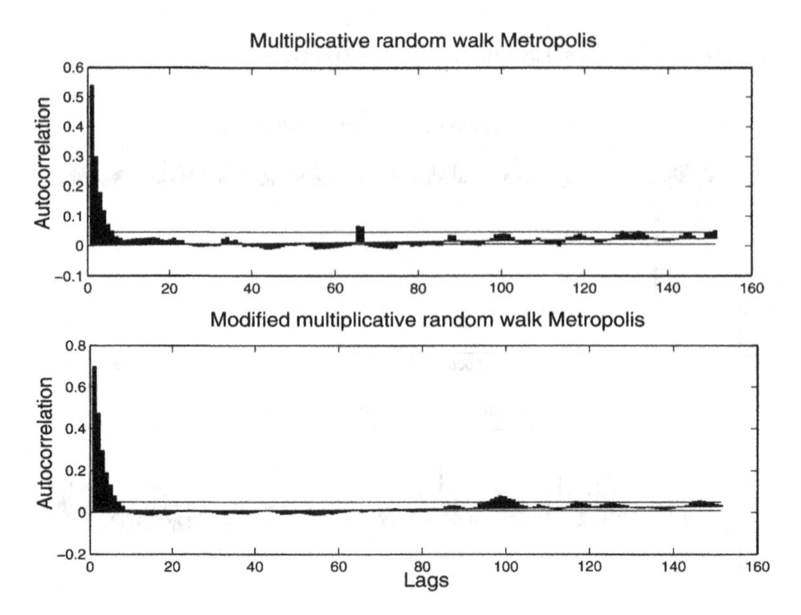

FIGURE 1.9. Autocorrelation functions of α for the multiplicative and the modified multiplicative Metropolis sampler

1.7.1 Introduction

A Bayesian approach to model selection proceeds as follows. Suppose that the data $\boldsymbol{y}$ are considered to have been generated by a model m, one of a set M of competing models. Each model specifies the distribution of $\boldsymbol{Y}$, $L(\boldsymbol{y}|m, \boldsymbol{\theta}_m)$ apart from an unknown parameter vector $\boldsymbol{\theta}_m \in \boldsymbol{\Theta}_m$, where $\boldsymbol{\Theta}_m$ is the set of all possible values for the parameters of model m. If $p(m)$ is the prior probability of model m, then the posterior probability is given by

$$\pi(m|\boldsymbol{y}) \quad = \quad \frac{p(m)L(\boldsymbol{y}|m)}{\sum\limits_{m \in M} p(m)L(\boldsymbol{y}|m)}, \qquad m \in M \tag{1.31}$$

where $L(\boldsymbol{y}|m)$ is the marginal likelihood calculated using

$$L(\boldsymbol{y}|m) = \int L(\boldsymbol{y}|m, \boldsymbol{\theta}_m) p(\boldsymbol{\theta}_m|m) d\boldsymbol{\theta}_m$$

and $p(\boldsymbol{\theta}_m|m)$ is the conditional prior distribution of $\boldsymbol{\theta}_m$, the model parameters for model m.

There are two major problems in calculating posterior model probabilities. First, calculation of the marginal likelihood meets the usual integration difficulties of Bayesian statistics. Second, when we deal with many models, calculation of $L(\boldsymbol{y}|m)$ for all $m \in M$ becomes infeasible. MCMC methods can be useful in both cases since they do not only provide methodologies for calculating the marginal likelihood from the MCMC output but they can also provide tools to construct *model search* algorithms which generate observations from the joint posterior distribution $\pi(m, \boldsymbol{\theta}_m|\boldsymbol{y})$ of $(m, \boldsymbol{\theta}_m)$. We shall deal with these two cases separately in the next two sections.

1.7.2 Marginal likelihood calculations

Together with the development of MCMC methods that estimate posterior expectations, there has been active research to exploit MCMC output to estimate $L(\boldsymbol{y}|m)$; see Newton & Raftery (1994), Gelfand & Dey (1994) and Kass & Raftery (1995). Since these methods require MCMC output from *each* model under consideration, their application is somewhat limited. Nevertheless, they are very important and rather easy to implement when only few models are considered so we shall describe here two methods that we think are important for spatial statistics: bridge sampling and path sampling.

Bridge sampling

Assume that we need to estimate the ratio of two marginal likelihoods for models m and m' of equal dimension, denoted by $L(\boldsymbol{y}|m)$ and $L(\boldsymbol{y}|m')$ respectively. If m and m' have different dimensions the methods of this section need extra care; see Chen & Shao (1997). To simplify notation, denote by $c_m(\boldsymbol{\theta})$ and $c_{m'}(\boldsymbol{\theta})$ the corresponding unnormalised posterior densities, so that $c_m(\boldsymbol{\theta}) = L(\boldsymbol{y}|\boldsymbol{\theta}_m, m)p(\boldsymbol{\theta}_m|m)$ and $c_{m'}(\boldsymbol{\theta}) = L(\boldsymbol{y}|\boldsymbol{\theta}_{m'}, m')p(\boldsymbol{\theta}_{m'}|m')$. A simple importance sampling estimator for the marginal likelihoods ratio is easily derived by noting that

$$\frac{L(\boldsymbol{y}|m)}{L(\boldsymbol{y}|m')} = \mathbf{E}_{m'}\left\{\frac{c_m(\boldsymbol{\theta})}{c_{m'}(\boldsymbol{\theta})}\right\} \tag{1.32}$$

where $\mathbf{E}_{m'}$ denotes the expected value with respect to $\pi(\boldsymbol{\theta}_{m'}|\boldsymbol{y}, m')$ and $\boldsymbol{\Theta}_m \subset \boldsymbol{\Theta}_{m'}$. Since this estimator uses draws from only one posterior density, it will produce poor results when the two posteriors have little overlap. Meng & Wong (1996) introduced *bridge sampling* by generalising the above equation so that the condition $\boldsymbol{\Theta}_m \subset \boldsymbol{\Theta}_{m'}$ is no longer necessary and draws from both posteriors are used:

$$\frac{L(\boldsymbol{y}|m)}{L(\boldsymbol{y}|m')} = \frac{\mathbf{E}_{m'}\{c_m(\boldsymbol{\theta})g(\boldsymbol{\theta})\}}{\mathbf{E}_m\{c_{m'}(\boldsymbol{\theta})g(\boldsymbol{\theta})\}} \tag{1.33}$$

where $g(\boldsymbol{\theta})$ is a positive function defined on $\boldsymbol{\Theta}_m \bigcap \boldsymbol{\Theta}_{m'}$ such that

$$0 < \left|\int_{\boldsymbol{\Theta}_m \bigcap \boldsymbol{\Theta}_{m'}} c_m(\boldsymbol{\theta})c_{m'}(\boldsymbol{\theta})g(\boldsymbol{\theta})d\boldsymbol{\theta}\right| < \infty. \tag{1.34}$$

Note that (1.33) can be used to estimate only $L(\boldsymbol{y}|m)$ if a known density is used instead of $\pi(\theta_{m'}|\boldsymbol{y}, m')$ so that $\pi(\theta_{m'}|\boldsymbol{y}, m') = c_{m'}(\boldsymbol{\theta})$ and $L(\boldsymbol{y}|m') = 1$.

Different choices of $g(\boldsymbol{\theta})$ produce different estimators. For example, assuming $\boldsymbol{\Theta}_m \subset \boldsymbol{\Theta}_{m'}$, $g(\boldsymbol{\theta}) = c_{m'}^{-1}(\boldsymbol{\theta})$ produces (1.32). If $\boldsymbol{\Theta}_m = \boldsymbol{\Theta}_{m'}$, $g(\boldsymbol{\theta}) = c_m^{-1}(\boldsymbol{\theta})c_{m'}^{-1}(\boldsymbol{\theta})$ produces a generalisation of the harmonic rule of Newton & Raftery (1994):

$$\frac{L(\boldsymbol{y}|m)}{L(\boldsymbol{y}|m')} = \frac{\mathbf{E}_{m'}\left\{c_{m'}^{-1}(\boldsymbol{\theta})\right\}}{\mathbf{E}_m\left\{c_m^{-1}(\boldsymbol{\theta})\right\}}. \tag{1.35}$$

Another interesting choice of $g(\boldsymbol{\theta})$ produces a generalisation of the marginal likelihood approach introduced by Chib (1995), which specifically relies on Metropolis-Hastings output; see Jeliazkov & Chib (2001).

Gelman & Meng (1998) studied bridge sampling in detail and they noted that the idea behind it is that samples are obtained from both posteriors whereas another density is served as a bridge to connect the two samples. To see this more clearly, assume that a 'bridge' density $\pi_b(\boldsymbol{\theta})$ lies 'between' $c_m(\boldsymbol{\theta})$ and $c_{m'}(\boldsymbol{\theta})$. Then $g(\boldsymbol{\theta})$ can be written as

$$g(\boldsymbol{\theta}) = \pi_b(\boldsymbol{\theta})/[c_m(\boldsymbol{\theta})c_{m'}(\boldsymbol{\theta})] \tag{1.36}$$

so that

$$\frac{L(\boldsymbol{y}|m)}{L(\boldsymbol{y}|m')} = \frac{\mathbf{E}_{m'}\left\{\pi_b(\boldsymbol{\theta})/c_{m'}(\boldsymbol{\theta})\right\}}{\mathbf{E}_m\left\{\pi_b(\boldsymbol{\theta})/c_m(\boldsymbol{\theta})\right\}}. \tag{1.37}$$

The message from the last equation is that the ratio of the two marginal likelihoods is estimated by using samples from both posterior densities of the two models with the density $\pi_b(\boldsymbol{\theta})$ served as a 'bridge' between them. Since the idea is based on the fact that this bridge is supported on $\boldsymbol{\Theta}_m \bigcap \boldsymbol{\Theta}_{m'}$ note that this method would not work if $\boldsymbol{\Theta}_m \bigcap \boldsymbol{\Theta}_{m'} = \emptyset$ and it would work poorly if this intersection is 'small'.

Path sampling

The natural extension to bridge sampling is to use a series of bridges which connect the two densities: this is achieved via *path sampling* (Gelman & Meng 1998). The method could be used to compare models without any restrictions on $\boldsymbol{\Theta}_m$ and $\boldsymbol{\Theta}_{m'}$. See also Sections 3.4.3 and 4.7.4.

Consider the family of posterior densities indexed by the parameter $z \in [0, 1]$:

$$\pi(\boldsymbol{\theta} \mid \boldsymbol{y}, z) = \frac{\pi(\boldsymbol{\theta}, \boldsymbol{y} \mid z)}{L(\boldsymbol{y} \mid z)} \tag{1.38}$$

and suppose that $z = 1$ corresponds to $L(\boldsymbol{y} \mid m)$ while $L(\boldsymbol{y} \mid m')$ is denoted by $z = 0$. The other z values can correspond to a genuine continuum of intermediate models or can be totally artificial with no statistical meaning at all.

First note that by assuming interchange of differentiation and integration, since $L(\boldsymbol{y} \mid z)$ is just the marginal density of $\pi(\boldsymbol{\theta}, \boldsymbol{y} \mid z)$, we obtain

$$\begin{aligned}\frac{d}{dz}\log L(\boldsymbol{y} \mid z) &= \int L^{-1}(\boldsymbol{y} \mid z)\frac{d}{dz}\pi(\boldsymbol{\theta}, \boldsymbol{y} \mid z)d\boldsymbol{\theta} \\ &= \mathbf{E}_{\boldsymbol{\theta}\mid\boldsymbol{y},z}\left[\frac{d}{dz}\log\pi(\boldsymbol{\theta}, \boldsymbol{y} \mid z)\right]. \qquad (1.39)\end{aligned}$$

If we treat z as a parameter with density $p(z)$ independent of $\boldsymbol{\theta}$, then $\frac{d}{dz}\log\pi(\boldsymbol{\theta}, \boldsymbol{y} \mid z) = \frac{d}{dz}\log L(\boldsymbol{y} \mid \boldsymbol{\theta}, z)$ and

$$\begin{aligned}\log\left(\frac{L(\boldsymbol{y} \mid z=1)}{L(\boldsymbol{y} \mid z=0)}\right) &= \int_0^1 \mathbf{E}_{\boldsymbol{\theta}\mid\boldsymbol{y},z}\left[\frac{d}{dz}\log L(\boldsymbol{y} \mid \boldsymbol{\theta}, z)\right]dz \\ &= \mathbf{E}_{\boldsymbol{\theta},z\mid\boldsymbol{y}}\left[\frac{\frac{d}{dz}\log L(\boldsymbol{y} \mid \boldsymbol{\theta}, z)}{p(z)}\right].\end{aligned}$$

The last quantity can be estimated with a random sample from $\pi(\boldsymbol{\theta}, z \mid \boldsymbol{y})$ of size n by

$$\log\left(\frac{L(\boldsymbol{y} \mid z=1)}{L(\boldsymbol{y} \mid z=0)}\right) \simeq n^{-1}\sum_{i=1}^{n}\frac{\frac{d}{dz}\log L(\boldsymbol{y} \mid \boldsymbol{\theta}_i, z_i)}{p(z_i)}. \qquad (1.40)$$

Gelman and Meng suggest calculating the integral over z with a trapezoidal rule based on a equally spaced set of points $z_0 = 0, z_1, z_2, \ldots, z_k = 1$ over $[0,1]$. The resulting estimator for samples $\boldsymbol{\theta}_{1,z_i}, \boldsymbol{\theta}_{2,z_i}, \ldots, \boldsymbol{\theta}_{n,z_i}$ from $\pi(\boldsymbol{\theta} \mid \boldsymbol{y}, z_i)$ for $i = 1, \ldots, k$ is

$$\begin{aligned}\log\left(\frac{L(\boldsymbol{y} \mid z=1)}{L(\boldsymbol{y} \mid z=0)}\right) &= \frac{1}{2}\sum_{i=0}^{k-1}(z_{i+1}-z_i)\left[\frac{1}{n}\sum_{t=1}^{n}\left(\left[\frac{\frac{d}{dz}\log\pi(\boldsymbol{y} \mid \boldsymbol{\theta}_t, z)}{p(z)}\right]_{z_{i+1}}\right.\right. \\ &+ \left.\left.\left[\frac{\frac{d}{dz}\log\pi(\boldsymbol{y} \mid \boldsymbol{\theta}_t, z)}{p(z)}\right]_{z_i}\right)\right].\end{aligned}$$

1.7.3 MCMC model-search selection methods

Independence sampler

The most straightforward MCMC approach from generating from the posterior distribution $\pi(m, \boldsymbol{\theta}_m|\boldsymbol{y})$ is a standard Metropolis-Hastings approach. Given the current value of $(m, \boldsymbol{\theta}_m)$, a proposal $(m', \boldsymbol{\theta}'_{m'})$ is generated from some proposal distribution over the parameter space for (m, Θ) and the proposal is accepted as the next observation of the chain with the usual Metropolis-Hastings acceptance probability given in Algorithm 1.3. This approach is considered in Gruet & Robert (1997) for mixture models. As usual, the independence sampler works best if the proposal q is a reasonable approximation to the target distribution f, so what is needed here is to approximate $\pi(m)$ and $\pi(\boldsymbol{\theta}_m|m)$ for every m; this is evidently very difficult to achieve.

Reversible jump

Instead of using an independence sampler to obtain samples from the density $\pi(m, \boldsymbol{\theta}_m|\boldsymbol{y})$, a sensible strategy would be of course to allow the proposal $(m', \boldsymbol{\theta}'_{m'})$ to depend on the current values $(m, \boldsymbol{\theta}_m)$. The standard Metropolis-Hastings algorithm (1.3) cannot be applied to a sample space of varying dimension. However, Green (1995) developed the reversible jump MCMC for exactly this situation. A nice introduction and review appears in Green (2002) whereas connections with other existing methods for model determination and variable selection via MCMC are described in Dellaportas, Forster & Ntzoufras (2002). The reversible jump approach for generating from $\pi(m, \boldsymbol{\theta}_m|\boldsymbol{y})$ is based on creating a Markov chain which can 'jump' between models with parameter spaces of different dimension in a flexible way, while retaining detailed balance which ensures the correct limiting distribution provided the chain is irreducible and aperiodic. It is a generalisation of ideas that first appeared in spatial statistics for Markov point processes, see Geyer & Møller (1994) and Section 4.7.7, and in image analysis, see Grenander & Miller (1994). The algorithm allows the calculation of posterior model probabilities for models with parameters of any dimension. Its major drawback is that the efficiency of the algorithm requires careful choice of proposal densities, though see Brooks, Giudici & Roberts (2002) for generic solutions to this problem. The algorithm proceeds as follows.

Suppose that the current state of the Markov chain is $(m, \boldsymbol{\theta}_m)$, where $\boldsymbol{\theta}_m$ has dimension $d(\boldsymbol{\theta}_m)$, then one version of the procedure is given by Algorithm 3 below.

There are many variations or simpler versions of reversible jump that can be applied in specific model determination problems. For example, if all parameters of the proposed model are generated directly from a proposal distribution, then $(\boldsymbol{\theta}'_{m'}, \boldsymbol{u}') = (\boldsymbol{u}, \boldsymbol{\theta}_m)$ with $d(\boldsymbol{\theta}_m) = d(\boldsymbol{u}')$ and $d(\boldsymbol{\theta}_{m'}) = d(\boldsymbol{u})$, and the Jacobian term in (3) is one. In this case, we have an independence sampler. Therefore the independence sampler is a special case of reversible jump. With the same proposals, but where the function $(\boldsymbol{\theta}'_{m'}, \boldsymbol{u}') = g_{m,m'}(\boldsymbol{u}, \boldsymbol{\theta}_m)$ is not the identity, we have a more general Metropolis-Hastings algorithm where $\boldsymbol{\theta}'_{m'}$ is allowed to depend on $\boldsymbol{\theta}_m$. If $m' = m$, then the move is a standard Metropolis-Hastings step.

However, the real flexibility of the reversible jump formulation is that it allows us to use proposal distributions of lower dimension than $d(\boldsymbol{\theta}'_{m'})$. For example, if model m is nested in m' then there may be an extremely natural proposal distribution and transformation function $g_{m,m'}$ (which may be the identity function) such that $d(\boldsymbol{u}') = 0$ and $\boldsymbol{\theta}'_{m'} = g_{m,m'}(\boldsymbol{\theta}_m, \boldsymbol{u})$.

Algorithm 3 *The reversible jump MCMC algorithm for a model selection problem.*

- `Propose a new model` m' `with probability` $j(m, m')$.
- `Generate` $\boldsymbol{u}$ `(which can be of lower dimension than` $\boldsymbol{\theta}_{m'}$`) from a specified proposal density` $q(\boldsymbol{u}|\boldsymbol{\theta}_m, m, m')$.
- `Set` $(\boldsymbol{\theta}'_{m'}, \boldsymbol{u}') = g_{m,m'}(\boldsymbol{\theta}_m, \boldsymbol{u})$ `where` $g_{m,m'}$ `is a specified invertible function. Hence` $d(\boldsymbol{\theta}_m) + d(\boldsymbol{u}) = d(\boldsymbol{\theta}_{m'}) + d(\boldsymbol{u}')$. `Note that` $g_{m',m} = g^{-1}_{m,m'}$.
- `Accept the proposed move to model` m' `with probability`

$$\alpha = \min\left(1, \frac{L(\boldsymbol{y}|m', \boldsymbol{\theta}'_{m'})p(\boldsymbol{\theta}'_{m'}|m')p(m')j(m', m), q(\boldsymbol{u}'|\boldsymbol{\theta}_m, m', m)}{L(\boldsymbol{y}|m, \boldsymbol{\theta}_m)p(\boldsymbol{\theta}_m|m)p(m)j(m, m')q(\boldsymbol{u}|\boldsymbol{\theta}_{m'}, m, m')}|J|\right)$$

`where`

$$J = \frac{\partial g_{m,m'}(\boldsymbol{\theta}_m, \boldsymbol{u})}{\partial(\boldsymbol{\theta}_m, \boldsymbol{u})}.$$

Therefore, when the reverse move is proposed, the model parameters are proposed deterministically. This is the approach taken, for example, in Dellaportas & Forster (1999). More interesting than the above case, and indeed the research challenge of reversible jump, is the construction of moves via appropriate choices of $g_{m,m'}$. Some applications are Richardson & Green (1997) for finite mixtures, Robert, Ryden & Titterington (2000) for hidden Markov models and Ntzoufras, Dellaportas & Forster (2002) for generalised linear models.

Another interesting situation arises when, for each m, the posterior density $\pi(\boldsymbol{\theta}_m|m, \boldsymbol{y})$ is available and can be integrated to give the marginal likelihood, $\pi(\boldsymbol{y}|m)$. If this distribution is used as a proposal in the reversible jump algorithm, then the acceptance probability is given by

$$\begin{aligned} \alpha &= \min\left(1, \frac{L(\boldsymbol{y}|m', \boldsymbol{\theta}'_{m'})p(\boldsymbol{\theta}'_{m'}|m')p(m')j(m', m)\pi(\boldsymbol{\theta}_m|m, \boldsymbol{y})}{L(\boldsymbol{y}|m, \boldsymbol{\theta}_m)p(\boldsymbol{\theta}_m|m)p(m)j(m, m')\pi(\boldsymbol{\theta}'_{m'}|m', \boldsymbol{y})}\right) \\ &= \min\left(1, B_{m'm}\frac{p(m')j(m', m)}{p(m)j(m, m')}\right) \end{aligned}$$

where $B_{m'm}$ is the Bayes factor of model m' against model m. In practice, we cannot usually calculate $B_{m'm}$. In the special case where models are decomposable graphical models, Madigan & York (1995) used exactly this approach, which they called MC^3. Moreover, if the marginal likelihood can be somehow easily approximated, for example by Laplace approximations as described in Kass & Raftery (1995), this algorithm provides a very fast search in the model space. Note that there is no need to generate the model parameters $\boldsymbol{\theta}_m$ as part of the Markov chain. These can be generated separately from the known posterior distributions $\pi(\boldsymbol{\theta}_m|m, \boldsymbol{y})$ if required.

There are a number of related approaches to MCMC algorithms which need to traverse different dimensional spaces. A very elegant and flexible formulation merely writes all densities with respect to a composite marked Poisson process measure, see for example Geyer & Møller (1994), Tierney (1998). In this context, natural proposal distributions present themselves in terms of birth and death and displacement samplers. These algorithms are particularly natural and useful in the context of the modelling of spatial point processes which are often easy to characterise in terms of their density with respect to a suitable Poisson process measure. This approach will be considered in detail in Chapter 4.

Acknowledgments: We would like to thank Loukia Meligotsidou, Omiros Papaspiliopoulos and Stefano Tonellato for useful comments and discussions.

1.8 References

Besag, J. (1994). Comments on "Representations of knowledge in complex systems" by U. Grenander and M.I. Miller, *Journal of the Royal Statistical Society, Series B* **56**: 591–592.

Besag, J. & Green, P. J. (1993). Spatial statistics and Bayesian computation (with discussion), *Journal of the Royal Statistical Society, Series B* **55**: 25–37.

Brooks, S. P., Giudici, P. & Roberts, G. O. (2002). Efficient construction of reversible jump MCMC proposal distributions, *Journal of the Royal Statistical Society, Series B* **64**. To appear.

Brooks, S. P. & Roberts, G. O. (1998). Diagnosing convergence of Markov chain Monte Carlo algorithms, *Statistics and Computing* **8**: 319–335.

Chen, M. & Shao, Q.-M. (1997). Estimating ratios of normalizing constants for densities with different dimensions, *Statistica Sinica* **7**: 607–630.

Chib, S. (1995). Marginal likelihood from the Gibbs output, *Journal of the American Statistical Association* **90**: 1313–1321.

Cowles, M. K. & Carlin, B. P. (1996). Markov chain Monte Carlo convergence diagnostics: a comparative review, *Journal of the American Statistical Association* **91**: 883–904.

Damien, P., Wakefield, J. & Walker, S. (1999). Gibbs sampling for Bayesian non-conjugate and hierarchical models by using auxiliary variables, *Journal of the Royal Statistical Society, Series B* **61**: 331–344.

Dellaportas, P. & Forster, J. J. (1999). Markov chain Monte Carlo model determination for hierarchical and graphical log-linear models, *Biometrika* **86**: 615–633.

Dellaportas, P., Forster, J. J. & Ntzoufras, I. (2002). On Bayesian model and variable selection using MCMC, *Statistics and Computing* **12**: 27–36.

Dellaportas, P. & Smith, A. F. M. (1993). Bayesian inference for generalised linear and proportional hazards models via Gibbs sampling, *Applied Statistics* **42**: 443–459.

Devroye, L. (1986). *Non-Uniform Random Variate Generation*, Springer-Verlag, New York.

Edwards, R. & Sokal, A. (1988). Generalization of the Fortium-Kasteleyn-Swendsen-Wang representation and Monte Carlo algorithm, *Physical Review Letters D* **38**: 2009–2012.

Gamerman, D. (1997). *Markov Chain Monte Carlo: Stochastic Simulation for Bayesian Inference*, Chapman and Hall, London.

Gelfand, A. E. & Dey, D. K. (1994). Bayesian model choice: Asymptotics and exact calculations, *Journal of the Royal Statistical Society, Series B* **56**: 501–514.

Gelfand, A. E. & Smith, A. F. M. (1990). Sampling-based approaches to calculating marginal densities, *Journal of the American Statistical Association* **85**: 398–409.

Gelman, A. & Meng, X. L. (1998). Simulating normalising constants: From importance sampling to bridge sampling to path sampling, *Statistical Science* **13**: 163–185.

Gelman, A., Roberts, G. O. & Gilks, W. (1996). Efficient Metropolis jumping rules, *Bayesian Statistics V* pp. 599–608.

Gelman, A. & Rubin, D. (1992). Inference from iterative simulation using multiple sequences, *Statistical Science* **7**: 457–472.

Geman, S. & Geman, D. (1984). Stochastic relaxation, Gibbs distributions and the Bayesian restoration of images, *IEEE Transactions of Pattern Analysis and Machine Intelligence* **6**: 721–741.

Geyer, C. (1992). Practical Markov chain Monte Carlo, *Statistical Science* **7**: 473–483.

Geyer, C. J. & Thompson, E. A. (1992). Constrained Monte Carlo maximum likelihood for dependent data (with discussion), *Journal of the Royal Statistical Society, Series B* **38**: 657–699.

Geyer, C. & Møller, J. (1994). Simulation procedures and likelihood inference for spatial point processes, *Scandinavian Journal of Statistics* **21**: 359–373.

Gilks, W. R., Richardson, S. & Spiegelhalter, D. J. (eds) (1996). *Markov Chain Monte Carlo in Practice*, Chapman and Hall, London.

Gilks, W. & Wild, P. (1992). Adaptive rejection sampling for Gibbs sampling, *Applied Statistics* **41**: 37–348.

Green, P. J. (1995). Reversible jump Markov chain Monte Carlo computation and Bayesian model determination, *Biometrika* **82**: 711–732.

Green, P. J. (2002). Reversible jump MCMC, *in* P. J. Green, N. Hjort & S. Richardson (eds), *Highly Structured Stochastic Systems*, Oxford University Press. To appear.

Grenander, U. & Miller, M. (1994). Representations of knowledge in complex systems (with discussion), *Journal of the Royal Statistical Society, Series B* **56**: 549–603.

Gruet, M.-A. & Robert, C. (1997). Comment on 'Bayesian analysis of mixtures with an unknown number of components' by S. Richardson and P.J. Green, *Journal of Royal Statistical Society Series B* **59**: 777.

Hills, S. E. & Smith, A. F. M. (1992). Parameterization issues in Bayesian inference, *in* J. Bernardo, J. Berger, A. Dawid & A. Smith (eds), *Bayesian Statistics 4*, Oxford University Press, Oxford, pp. 227–246.

Jarner, S. & Roberts, G. O. (2002a). Convergence of heavy tailed MCMC algorithms, *submitted for publication, available at* `http://www.statslab.cam.ac.uk/~mcmc` .

Jarner, S. & Roberts, G. O. (2002b). Polynomial convergence rates of Markov chains, *Annals of Applied Probability* . To appear.

Jeliazkov, I. & Chib, S. (2001). Marginal likelihood from the Metropolis-Hastings output, *Journal of the American Statistical Association* **96**: 270–281.

Kass, R. E., Carlin, B. P., Gelman, A. & Neal, R. (1998). Markov chain Monte Carlo in practice: a roundtable discussion, *The American Statistician* **52**: 93–100.

Kass, R. & Raftery, A. (1995). Bayes factors, *Journal of the American Statistical Association* **90**: 773–795.

Liu, J. (2001). *Monte Carlo Strategies in Scientific Computing*, Springer-Verlag, New York.

Madigan, D. & York, J. (1995). Bayesian graphical models for discrete data, *International Statistical Review* **63**: 215–232.

Meng, X. L. & van Dyk, D. A. (1997). The EM algorithm - an old folk-song to a fast new tune, *Journal of the Royal Statistical Society, Series B* **59**: 511–567.

Meng, X. L. & Wong, W. H. (1996). Simulating ratios of normalising constants via a simple identity: A theoretical exploration, *Statistica Sinica* **6**: 831–860.

Mengersen, K. L., Robert, C. P. & Guihenneuc-Jouyaux, C. (1999). MCMC convergence diagnostics: A review, *in* A. P. D. J. M. Bernardo, J. O. Berger & A. F. M. Smith (eds), *Bayesian Statistics 6*, Oxford University Press, UK, pp. 415–440.

Metropolis, N., Rosenbluth, A., Rosenbluth, M., Teller, A. & Teller, E. (1953). Equations of state calculations by fast computing machines, *Journal of Chemical Physics* **21**: 1087–1091.

Meyn, S. P. & Tweedie, R. L. (1993). *Markov Chains and Stochastic Stability*, Springer-Verlag, London.

Møller, J. (1999). Markov chain Monte Carlo and spatial point processes, *in* O. E. Barndorff-Nielsen, W. S. Kendall & M. N. M. van Lieshout (eds), *Stochastic Geometry: Likelihood and Computations*, number 80 in *Monographs on Statistics and Applied Probability*, Chapman and Hall/CRC, Boca Raton, pp. 141–172.

Neal, R. M. (2002). Slice sampling, with discussion, *Annals of Statistics* . To appear.

Newton, M. A. & Raftery, A. E. (1994). Approximate Bayesian inference by the weighted likelihood bootstrap (with discussion), *Journal of the Royal Statistical Society, Series B* **56**: 1–48.

Ntzoufras, I., Dellaportas, P. & Forster, J. (2002). Bayesian variable and link determination for generalised linear models, *Journal of Statistical Planning and Inference* . To appear.

Papaspiliopoulos, O., Roberts, G. O. & Skold, M. (2002). Parameterisation of hierarchical models and data augmentation, *in* J. M. Bernardo, J. Berger, A. P. Dawid & A. F. M. Smith (eds), *Bayesian Statistics 7*, Oxford University Press, UK. To appear.

Richardson, S. & Green, P. J. (1997). On Bayesian analysis of mixtures with an unknown number of components (with discussion), *Journal of the Royal Statistical Society, Series B* **59**: 731–792.

Robert, C. P. & Casella, G. (1999). *Monte Carlo Statistical Methods*, Springer, New York.

Robert, C., Ryden, T. & Titterington, D. (2000). Bayesian inference in hidden Markov models through the reversible jump Markov chain Monte Carlo method, *Journal of the Royal Statistical Society, Series B* **62**: 57–75.

Roberts, G. O. (2002). Linking theory and practice of MCMC, *in* P. J. Green, N. Hjort & S. Richardson (eds), *Highly Structured Stochastic Systems*, Oxford University Press. To appear.

Roberts, G. O., Gelman, A. & Gilks, W. (1997). Weak convergence and optimal scaling of random walk Metropolis algorithms, *Annals of Applied Probability* **7**: 110–120.

Roberts, G. O. & Rosenthal, J. S. (1997). Geometric ergodicity and hybrid Markov chains, *Electronic Communications in Probability* **2**: 13–25.

Roberts, G. O. & Rosenthal, J. S. (2001a). Markov chains and de-initialising processes, *Scandinavian Journal of Statistics* **28**: 489–504.

Roberts, G. O. & Rosenthal, J. S. (2001b). Optimal scaling of various Metropolis-Hastings algorithms, *Statistical Science* **16**: 351–367.

Roberts, G. O. & Sahu, S. K. (1997). Updating schemes, correlation structure, blocking and parameterisation for the Gibbs sampler, *Journal of the Royal Statistical Society, Series B* **59**: 291–397.

Roberts, G. O. & Skold, M. (2001). Marginal density estimation for the Metropolis-Hastings algorithm, *available at* `http://www.maths.lancs.ac.uk/~robertgo/RobSko01.ps` .

Roberts, G. O. & Smith, A. (1994). Simple conditions for the convergence of the Gibbs sampler and Metropolis-Hastings algorithms, *Stochastic Processes and their Applications* **49**: 207–216.

Roberts, G. O. & Tweedie, R. L. (1996). Exponential convergence of Langevin diffusions and their discrete approximations, *Bernoulli* **2**: 341–364.

Roberts, G. O. & Tweedie, R. L. (1998). Optimal scaling of discrete approximations to Langevin diffusions, *Journal of the Royal Statistical Society, Series B* **60**: 255–268.

Roberts, G. O. & Tweedie, R. L. (2002). *Understanding MCMC*, in preparation.

Rue, H. (2001). Fast sampling of Gaussian Markov random fields, *Journal of the Royal Statistical Society, Series B* **63**: 325–338.

Smith, A. F. M. & Roberts, G. O. (1993). Bayesian computation via the Gibbs sampler and related Markov chain Monte Carlo methods (with discussion), *Journal of the Royal Statistical Society, Series B* **55**: 3–24.

Stramer, O. & Tweedie, R. L. (1999). Langevin-Type Models II: Self-targeting candidates for MCMC algorithms, *Methodology and Computing in Applied Probability* pp. 307–328.

Suomela, P. (1976). Construction of nearest neighbour systems, *Dissertation 10, Department of Mathematics, University of Helsinki* .

Swendsen, R. H. & Wang, J. S. (1987). Non-universal critical dynamics in Monte Carlo simulations, *Physical Review Letters* **58**: 86–88.

Tanner, M. (1996). *Tools for Statistical Inference: Methods for Exploration of Posterior Distributions and Likelihood Functions, 3rd edition*, Springer-Verlag, New York.

Tierney, L. (1994). Markov chains for exploring posterior distributions (with discussion), *Annals of Statistics* **22**: 1701–1762.

Tierney, L. (1998). A note on Metropolis-Hastings kernels for general state-spaces, *Annals of Applied Probability* **8**: 1–9.

2

An Introduction to Model-Based Geostatistics

Peter J. Diggle
Paulo J. Ribeiro Jr.
Ole F. Christensen

2.1 Introduction

The term *geostatistics* identifies the part of spatial statistics which is concerned with continuous spatial variation, in the following sense. The scientific focus is to study a spatial phenomenon, $s(x)$ say, which exists throughout a continuous spatial region $A \subset \mathbb{R}^2$ and can be treated as if it were a realisation of a stochastic process $S(\cdot) = \{S(x) : x \in A\}$. In general, $S(\cdot)$ is not directly observable. Instead, the available data consist of measurements $Y_1, \ldots, Y_n$ taken at locations $x_1, \ldots, x_n$ sampled within A, and Y_i is a noisy version of $S(x_i)$. We shall assume either that the sampling design for $x_1, \ldots, x_n$ is deterministic or that it is stochastic but independent of the process $S(\cdot)$, and all analyses are carried out conditionally on $x_1, \ldots, x_n$.

The subject has its origins in problems connected with estimation of ore reserves in the mining industry (Krige 1951). Its subsequent development by Matheron and colleagues at École des Mines, Fontainebleau took place largely independently of "mainstream" spatial statistics. Standard references to this "classical" approach to geostatistics include Journel & Huijbregts (1978) and Chilès & Delfiner (1999). Parallel developments by Matérn (1960) and Whittle (1954, 1962, 1963) eventually led to the integration of classical geostatistics within spatial statistics. For example, Ripley (1981) re-cast the common geostatistical technique known as kriging within the framework of stochastic process prediction, whilst Cressie (1993) identified geostatistics as one of the three main sub-branches of spatial statistics. Significant cross-fertilisation continued throughout the 1980's and 1990's, but there is still vigorous debate on practical issues, such as the need (or not) for different approaches to prediction and parameter estimation, and the role of explicit probability models. The term *model-based geostatistics* was coined by Diggle, Tawn & Moyeed (1998) to mean the application of explicit parametric stochastic models and formal, likelihood-based methods of inference to geostatistical problems.

Our goal in this chapter is to introduce the reader to the model-based approach, in the sense intended by Diggle et al. (1998). We first describe two motivating examples, and formulate the general modelling framework for geostatistical problems, emphasising the key role of spatial prediction within the general framework. We then investigate the widely used special case of the Gaussian model, and discuss both maximum likelihood and Bayesian methods of inference. We present the results from an illustrative case-study based on one of our two motivating examples. We then consider non-Gaussian models, with a particular focus on generalised linear spatial models. The chapter concludes with some discussion, information on software and further references.

2.2 Examples of geostatistical problems

2.2.1 Swiss rainfall data

This is a standard data-set which has been widely used for empirical comparison of different methods of spatial interpolation (further information can be found at *ftp://ftp.geog.uwo.ca/SIC97*). The scientific problem posed by the data is to construct a continuous spatial map of rainfall values from observed values at a discrete set of locations. The original data consist of rainfall measurements on 8 May 1986 from 467 locations in Switzerland. The convention adopted in earlier analyses of these data is to use 100 of the data-points, as shown in Figure 2.1, to formulate and fit models to the data, and for prediction at locations without observations, whilst reserving the remaining 367 for empirical validation of the resulting predictions. In our illustrative analysis reported in Section 2.8 we use only the first 100 data points.

2.2.2 Residual contamination of Rongelap Island

These data are from a study of residual contamination on a Pacific island, Rongelap, following the USA's nuclear weapons testing programme during the 1950's (Diggle, Harper & Simon 1997). The island was evacuated in 1985, and a large, multi-disciplinary study was subsequently undertaken to determine whether the island is now safe for re-settlement. Within this overall objective, a specific goal was to estimate the spatial variation in residual contamination over the island, with a particular interest in the maximum level of contamination. To this end, a survey was carried out and noisy measurements Y_i of radioactive caesium concentrations were obtained initially on a grid of locations x_i at 200m spacing which was later supplemented by in-fill squares at 40m spacing. Figure 2.2 shows a map of the sampling locations $x_1, \ldots, x_{157}$. The in-fill squares are particularly useful for identifying and fitting a suitable model to the data, because they give direct information about the small-scale spatial correlation structure.

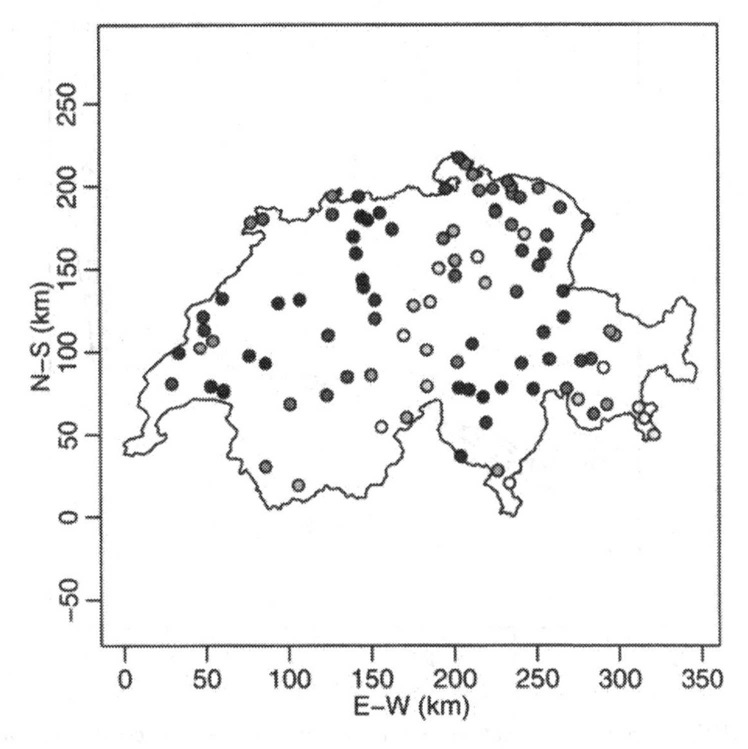

FIGURE 2.1. Swiss rainfall data at sample locations $x_1, \ldots, x_{100}$. Grey scale from white (low values) to black (high values) corresponds to the values of the observed rainfall, $y_1, \ldots, y_{100}$.

Design issues of this kind can have an important effect on the efficiency of any subsequent inferences. Generally speaking, placing sampling locations in a regular grid to cover the study region would be efficient for spatial prediction if all model parameters were known, whereas deliberately including more closely spaced sub-sets of sampling locations leads to more efficient estimation of certain model parameters. In this introductory account we shall not discuss design issues further.

A full analysis of the Rongelap island data, using a spatial Poisson log-linear model and Bayesian inference, is given in Diggle et al. (1998).

2.3 The general geostatistical model

We shall adopt the following general model and notation. Firstly, the data for analysis are of the form $(x_i, y_i) : i = 1, ..., n$, where $x_1, \ldots, x_n$ are locations within a study region $A \subset \mathbb{R}^2$ and $y_1, \ldots, y_n$ are measurements associated with these locations. We call $\{x_i : i = 1, ..., n\}$ the *sampling design* and assume that y_i is a realisation of $Y_i = Y(x_i)$, where $Y(\cdot) =$

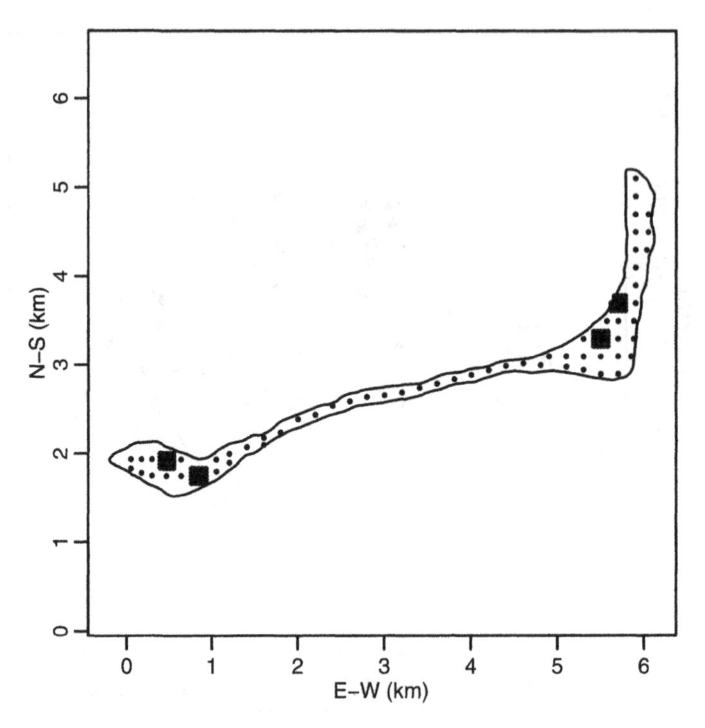

FIGURE 2.2. Sampling locations for the survey of residual contamination on Rongelap Island.

$\{Y(x) : x \in A\}$ is the *measurement process*. We postulate the existence of an unobserved stochastic process $S(\cdot) = \{S(x) : x \in A\}$, called the *signal process*; often in practice, $Y_1, \dots, Y_n$ are noisy versions of $S(x_1), \dots, S(x_n)$. Prediction is an integral part of a geostatistical analysis. We call $T = T(S(\cdot))$ the *target for prediction.* A *geostatistical model* is a specification of the joint distribution of the measurement process and the signal process, of the form $[S(\cdot), Y(\cdot)] = [Y(\cdot)|S(\cdot)][S(\cdot)]$, where $[\cdot]$ means "the distribution of." Note in particular that this model does not specify the distribution of the sampling design, which as noted earlier is assumed to be independent of both $S(\cdot)$ and $Y(\cdot)$. A *predictor* of T is any function $\hat{T} = \hat{T}(Y)$ where $Y = (Y_1, \dots, Y_n)^{\mathrm{T}}$. The *minimum mean square error predictor* minimises $MSE(\hat{T}) = \mathrm{E}[(T - \hat{T})^2]$, where the expectation is taken with respect to the joint distribution of T and Y. We have the following general result.

Proposition 1 *Provided that* $\mathrm{Var}[T] < \infty$, *the minimum mean square error predictor of* T *is* $\hat{T} = \mathrm{E}_T[T|Y]$, *with associated prediction mean square error* $\mathrm{E}[(T - \hat{T})^2] = \mathrm{E}_Y \mathrm{Var}_T[T|Y]$.

It is easy to show that $\mathrm{E}[(T - \hat{T})^2] \leq \mathrm{Var}[T]$, with equality if T and Y are independent random variables.

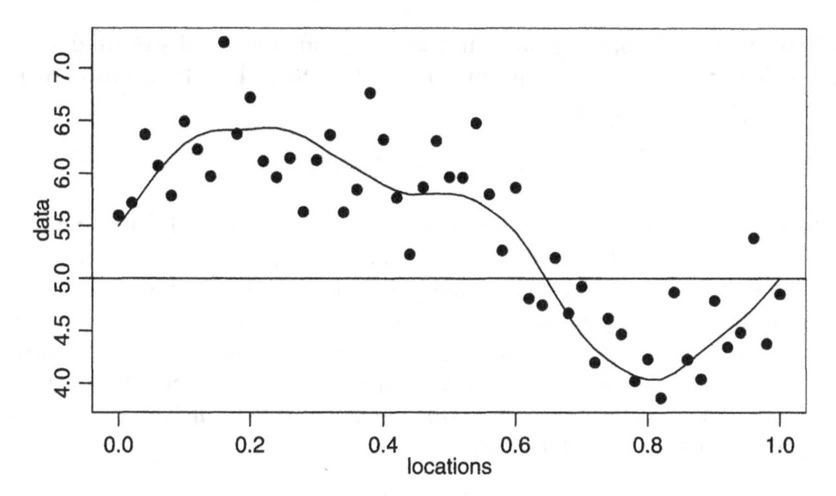

FIGURE 2.3. A simulation of the Gaussian model, illustrating the data $Y_1, \ldots, Y_n$ (dots), the signal $S(\cdot)$ (smooth curve) and the mean μ (horizontal line).

For point prediction, it is common practice to use $\mathrm{E}[T|y]$, the minimum mean square error predictor evaluated at the observed y. Similarly, for an estimate of the achieved mean square error, we would use the value of the prediction mean square error at the observed y, also called the *prediction variance*, $\mathrm{Var}[T|y]$. However, the complete answer to a prediction problem should be expressed as a probability distribution, $[T|y]$, called the *predictive distribution*. Within the Bayesian inferential paradigm which we shall eventually adopt, the predictive distribution coincides with the *posterior distribution* of T. From this point of view, the mean and variance of this posterior distribution are just two of many possible summary statistics. In particular, the mean is not transformation invariant; if $\hat{T}$ is the best predictor for T (in a mean square sense), this does not necessarily imply that $g(\hat{T})$ is the best predictor for $g(T)$.

2.4 The Gaussian Model

In the basic form of the *Gaussian* geostatistical model, $S(\cdot)$ is a stationary Gaussian process with $\mathrm{E}[S(x)] = \mu$, $\mathrm{Var}[S(x)] = \sigma^2$ and correlation function $\rho(u) = \mathrm{Corr}[S(x), S(x')]$, where $u = \|x - x'\|$, the Euclidean distance between x and x'. Also, the conditional distribution of Y_i given $S(\cdot)$ is Gaussian with mean $S(x_i)$ and variance τ^2, and $Y_i : i = 1, ..., n$ are mutually independent, conditional on $S(\cdot)$. Figure 2.3 shows a simulation of this model in one spatial dimension.

An equivalent formulation is that

$$Y_i = S(x_i) + Z_i : i = 1, ..., n,$$

where $Z_1, \ldots, Z_n$ are mutually independent, identically distributed with $Z_i \sim N(0, \tau^2), i = 1, ..., n$, and independent of $S(\cdot)$. The distribution of Y is multivariate Gaussian,

$$Y \sim N(\mu \mathbf{1}, \sigma^2 R + \tau^2 I)$$

where $\mathbf{1}$ denotes an n-element vector of ones, I is the $n \times n$ identity matrix and R is the $n \times n$ matrix with $(i,j)^{th}$ element $\rho(u_{ij})$ where $u_{ij} = ||x_i - x_j||$.

The specification of the correlation function, $\rho(u)$, determines the smoothness of the resulting process $S(\cdot)$. A formal mathematical description of the smoothness of a spatial surface $S(\cdot)$ is its degree of differentiability. A process $S(\cdot)$ is *mean-square continuous* if, for all x, $\mathrm{E}[\{S(x) - S(x')\}^2] \to 0$ as $\|x - x'\| \to 0$. Similarly, $S(x)$ is *mean square differentiable* if there exists a process $S'(\cdot)$ such that, for all x,

$$\mathrm{E}\left[\left\{\frac{S(x) - S(x')}{\|x - x'\|} - S'(x)\right\}^2\right] \to 0 \text{ as } \|x - x'\| \to 0.$$

The mean-square differentiability of $S(\cdot)$ is directly linked to the differentiability of its covariance function, according to the following result, a proof of which can be found in Chapter 2.4 in Stein (1999) or Chapter 5.2 in Cramér & Leadbetter (1967).

Proposition 2 *Let $S(\cdot)$ be a stationary Gaussian process with correlation function $\rho(u) : u \in \mathbb{R}$. Then, $S(\cdot)$ is mean-square continuous if and only if $\rho(u)$ is continuous at $u = 0$; $S(\cdot)$ is k times mean-square differentiable if and only if $\rho(u)$ is at least $2k$ times differentiable at $u = 0$.*

In general, continuity and/or differentiability in mean square do not imply the corresponding properties for realisations. However, within the Gaussian framework continuity or differentiability of realisations can be achieved by imposing slightly more strict smoothness conditions on the correlation function. For details, see Chapter 9 in Cramér & Leadbetter (1967), Adler (1981) and Kent (1989).

Amongst the various families of correlation function which have been proposed, the *Matérn* family is particularly attractive. Its algebraic form is given by

$$\rho(u) = \{2^{\kappa-1}\Gamma(\kappa)\}^{-1}(u/\phi)^{\kappa}K_{\kappa}(u/\phi)$$

where $\kappa > 0$ and $\phi > 0$ are parameters, and $K_\kappa(\cdot)$ denotes a Bessel function of order κ. Special cases include the *exponential* correlation function, $\rho(u) = \exp(-u/\phi)$, when $\kappa = 0.5$, and the *squared exponential* or *Gaussian* correlation function, $\rho(u) = \exp(-(u/\tilde{\phi})^2)$, when $\phi = \tilde{\phi}/(2\sqrt{\kappa + 1})$ and $\kappa \to \infty$. What makes the family particularly attractive is that the corresponding process $S(\cdot)$ is mean-square $\lceil\kappa - 1$ times differentiable where $\lceil\kappa$ denotes the largest integer less or equal to κ. Hence κ, which can be difficult

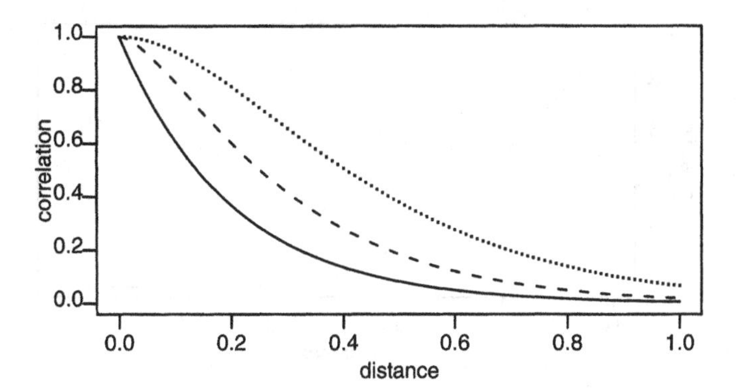

FIGURE 2.4. The Matérn correlation function with $\phi = 0.2$ and $\kappa = 1$ (solid line), $\kappa = 1.5$ (dashed line) and $\kappa = 2$ (dotted line).

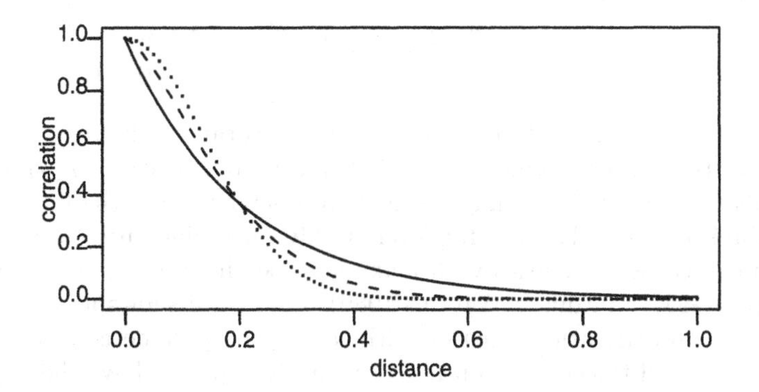

FIGURE 2.5. The powered exponential correlation function with $\phi = 0.2$ and $\kappa = 1$ (solid line), $\kappa = 1.5$ (dashed line) and $\kappa = 2$ (dotted line).

to estimate from noisy data, can be chosen to reflect scientific knowledge about the smoothness of the underlying process which $S(\cdot)$ is intended to represent. Figure 2.4 shows examples of the Matérn correlation function for $\kappa = 1, 1.5$ and 2.

Other families include the *powered exponential*,

$$\rho(u) = \exp\{-(u/\phi)^\kappa\},$$

defined for $\phi > 0$ and $0 < \kappa \leq 2$. This is less flexible than it first appears, because the corresponding process $S(\cdot)$ is mean-square continuous (but non-differentiable) if $\kappa < 2$, but mean-square infinitely differentiable if $\kappa = 2$, in which case the correlation matrix R may be very ill-conditioned. Figure 2.5 shows three examples of the powered exponential correlation function.

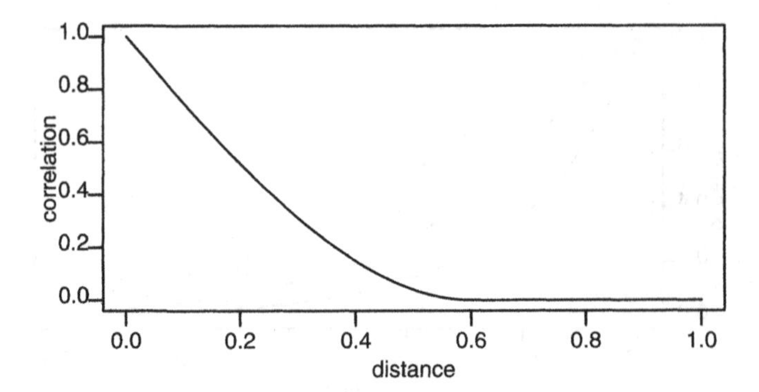

FIGURE 2.6. The spherical correlation function with $\phi = 0.6$.

In classical geostatistics, the *spherical* family is widely used. This has

$$\rho(u;\phi) = \begin{cases} 1 - \frac{3}{2}(u/\phi) + \frac{1}{2}(u/\phi)^3 & : \quad 0 \le u \le \phi \\ 0 & : \quad u > \phi \end{cases}$$

where $\phi > 0$ is a single parameter. One qualitative difference between this and the earlier families is that it has a finite range, i.e. $\rho(u) = 0$ for sufficiently large u. With only a single parameter it lacks the flexibility of the Matérn class. Also, the function is only once differentiable at $u = \phi$ which can cause difficulties with maximum likelihood estimation (Warnes & Ripley 1987, Mardia & Watkins 1989). Figure 2.6 shows an example of the spherical correlation function with correlation parameter $\phi = 0.6$.

Note that all the correlation functions presented here have the property that $\rho(u;\phi) = \rho_0(u/\phi)$; i.e. ϕ is a scale parameter with units of distance.

It is instructive to compare realisations of Gaussian processes with different correlation functions. For example, Figure 2.7 shows realisations of three different processes within the Matérn class, all generated from the same random number stream; the differences in smoothness as κ varies are very clear.

2.4.1 Prediction Under The Gaussian Model

Assume initially that the target for prediction is $T = S(x_0)$, the value of the signal process at a particular location x_0, where x_0 is not necessarily included within the sampling design. Under the Gaussian model, $[T, Y]$ is multivariate Gaussian. Therefore, $\hat{T} = \mathrm{E}[T|Y]$, the prediction variance $\mathrm{Var}[T|Y]$ and the predictive distribution $[T|Y]$ can be easily derived from the following standard result.

Proposition 3 *Let $X = (X_1, X_2)$ be multivariate Gaussian, with mean*

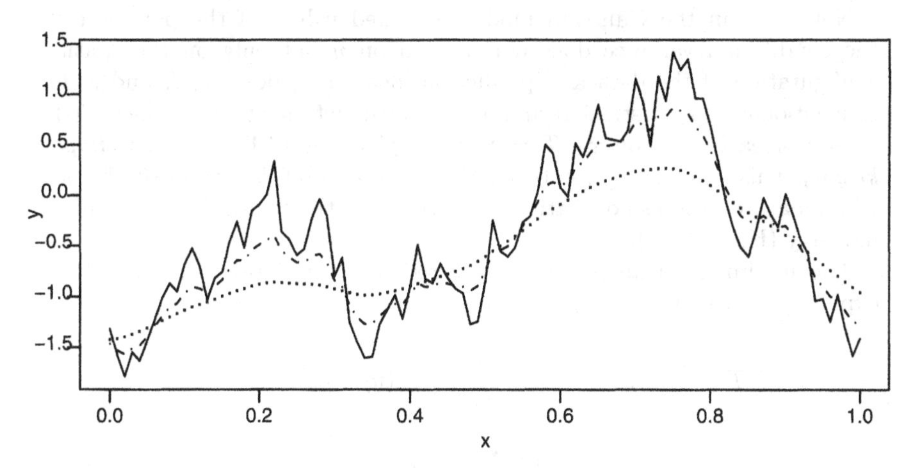

FIGURE 2.7. Simulations of Gaussian processes with Matérn correlation functions, using $\phi = 0.2$ and $\kappa = 0.5$ (solid line), $\kappa = 1$ (dashed line) or $\kappa = 2$ (dotted line).

vector $\mu = (\mu_1, \mu_2)$ *and covariance matrix*

$$\Sigma = \begin{bmatrix} \Sigma_{11} & \Sigma_{12} \\ \Sigma_{21} & \Sigma_{22} \end{bmatrix},$$

i.e. $X \sim N(\mu, \Sigma)$. *Then, the conditional distribution of* X_1 *given* $X_2 = x_2$ *is also multivariate Gaussian,* $X_1|X_2 = x_2 \sim N(\mu_{1|2}, \Sigma_{1|2})$, *where*

$$\mu_{1|2} = \mu_1 + \Sigma_{12}\Sigma_{22}^{-1}(x_2 - \mu_2)$$

and

$$\Sigma_{1|2} = \Sigma_{11} - \Sigma_{12}\Sigma_{22}^{-1}\Sigma_{21}.$$

For the geostatistical model $[T, Y]$ is multivariate Gaussian with mean vector $\mu\mathbf{1}$ and variance matrix

$$\begin{bmatrix} \sigma^2 & \sigma^2\mathbf{r}^{\mathrm{T}} \\ \sigma^2\mathbf{r} & \tau^2 I + \sigma^2 R \end{bmatrix}$$

where $\mathbf{r}$ is a vector with elements $r_i = \rho(||x_0 - x_i||) : i = 1, ..., n$. Hence, using Proposition 3 with $X_1 = T$ and $X_2 = Y$, we find that the minimum mean square error predictor for $T = S(x_0)$ is

$$\hat{T} = \mu + \sigma^2\mathbf{r}^{\mathrm{T}}(\tau^2 I + \sigma^2 R)^{-1}(y - \mu\mathbf{1}) \tag{2.1}$$

with prediction variance

$$\mathrm{Var}[T|y] = \sigma^2 - \sigma^2\mathbf{r}^{\mathrm{T}}(\tau^2 I + \sigma^2 R)^{-1}\sigma^2\mathbf{r}. \tag{2.2}$$

Note that in the Gaussian model, for fixed values of the parameters, the conditional variance does not depend on y but only on the spatial configuration of the data and prediction location(s) defining R and r. In conventional geostatistical terminology, construction of the surface $\hat{S}(\cdot)$, where for each location x_0, $\hat{T} = \hat{S}(x_0)$ is given by (2.1), is called *simple kriging*. This name was given by G. Matheron as a reference to D.G. Krige, who pioneered the use of statistical methods in the South African mining industry (Krige 1951).

The minimum mean square error predictor for $S(x_0)$ can be written explicitly as a linear function of the data y

$$
\begin{aligned}
\hat{T} = \hat{S}(x_0) &= \mu + \sum_{i=1}^{n} w_i(x_0)(y_i - \mu) \\
&= \{1 - \sum_{i=1}^{n} w_i(x_0)\}\mu + \sum_{i=1}^{n} w_i(x_0) y_i.
\end{aligned}
$$

Thus, the predictor $\hat{S}(x_0)$ compromises between its unconditional mean μ and the observed data y, the nature of the compromise depending on the target location x_0, the data-locations $x_1, \ldots, x_n$ and the values of the model parameters. We call $w_1(x_0), \ldots, w_n(x_0)$ the *prediction weights*. In general, the weight $w_i(x_0)$ tends to be large when x_i is close to x_0, $i = 1, \ldots, n$, and conversely, but this depends on the precise interplay between the sampling design and the assumed covariance structure of the data; in particular, even when the assumed correlation function is decreasing in distance, there is no guarantee that the weights will decrease with distance. Nor are they guaranteed to be positive, although in most practical situations large negative weights are rare.

One way to gain insight into the behaviour of the simple kriging predictor, $\hat{S}(\cdot)$, is to compute it for particular configurations of data under a range of assumed covariance structures. Note in particular the following general features of $\hat{S}(\cdot)$. Firstly, the surface $\hat{S}(\cdot)$ interpolates the data (meaning that $\hat{S}(x_i) = y_i$ for all x_i in the sampling design) if and only if $\tau^2 = 0$, since in this case $Y(x_i) = S(x_i)$ for $i = 1, ..., n$. When $\tau^2 > 0$, $\hat{S}(\cdot)$ tends to smooth out extreme fluctuations in the data. Secondly, for the correlation models considered here, $\hat{S}(\cdot)$ inherits the analytic smoothness at the origin of the assumed correlation function of $S(\cdot)$. So, for example, within the Matérn class, $\kappa \leq 0.5$ leads to a continuous but non-differentiable surface $\hat{S}(\cdot)$ whereas $\kappa > 0.5$ produces a smoother, differentiable surface. Finally, for typical correlation models in which $\rho(u) \to 0$ as $u \to \infty$, $\hat{S}(x_0) \approx \mu$ for a location x_0 sufficiently remote from all x_i in the sampling design, whereas when x_0 is close to one or more x_i, the corresponding $\hat{S}(x_0)$ will be more strongly influenced by the y_i's at these adjacent sampling locations.

Figure 2.8 illustrates some of these points, in the case of a small, one-dimensional data-set. The lines in the upper panel are the point predictions

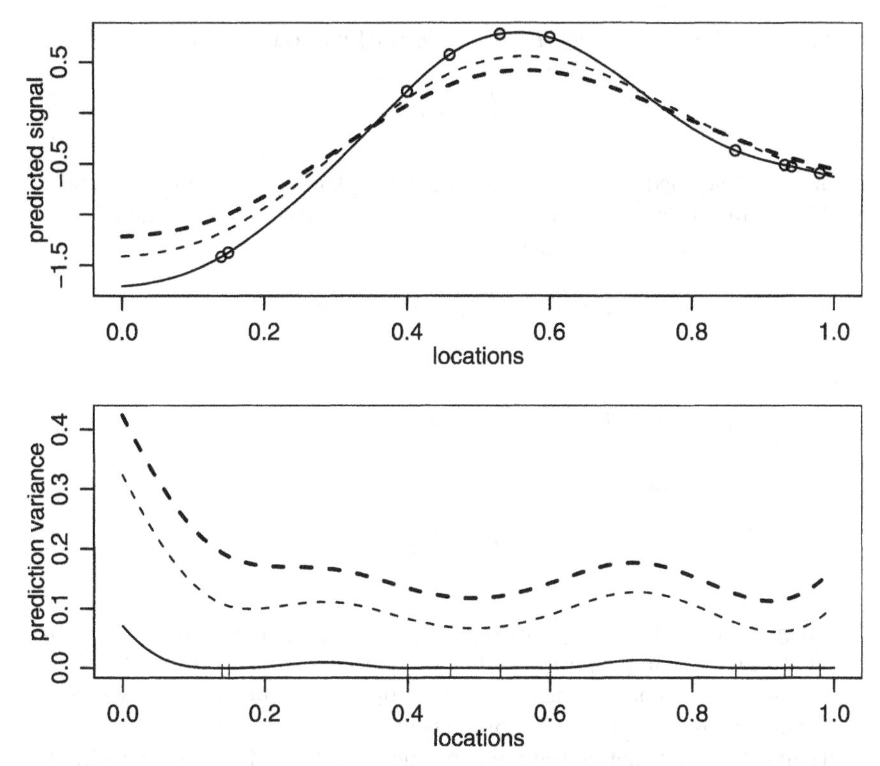

FIGURE 2.8. Point predictions and the data indicated by circles (upper panel) and prediction variances (lower panel) from 10 randomly spaced sampling locations indicated by the tick-marks in lower panel, assuming a Matérn correlation function with $\phi = 0.2$ and $\kappa = 2$, $\sigma^2 = 1$ and τ^2: 0 (solid line), 0.25 (thin dashed line) and 0.5 (thick dashed line).

$\hat{S}(x), x \in [0; 1]$ obtained using the data indicated by the circles. The data y are assumed to follow the model $Y_i = S(x_i) + Z_i$ where $S(\cdot)$ has mean $\mu = 0$, signal variance $\sigma^2 = 1$ and a Matérn correlation function with $\phi = 0.2$ and $\kappa = 2$, and Z_i are mutually independent with zero mean and variance τ^2, and the Z_i are independent of $S(\cdot)$. Holding the data fixed, Figure 2.8 shows the predictions which result when we assume each of $\tau^2 = 0$, 0.25 and 0.5. We observe that at data locations, when $\tau^2 = 0$ the predicted values coincide with the data. The higher the value of τ^2 the more the predictions approach the overall mean. The lower panel shows the corresponding prediction variances with tick-marks indicating the data locations.

In many applications, the inferential focus is not on $S(x_0)$ at a specific location x_0, but on some other property which can be expressed as a functional of the complete surface $S(\cdot)$, for example an areal average or maximum value. Firstly, let T be any *linear* functional of $S(\cdot)$,

$$T = \int_A w(x)S(x)dx$$

for some prescribed weighting function $w(x)$. Under the Gaussian model, $[T, Y]$ is multivariate Gaussian, hence $[T|y]$ is univariate Gaussian and the conditional mean and variance are

$$\mathrm{E}[T|y] = \int_A w(x)\mathrm{E}[S(x)|y]dx$$

and

$$\mathrm{Var}[T|y] = \int_A \int_A w(x)w(x')\mathrm{Cov}[S(x), S(x') \mid y]dxdx'.$$

Note in particular that

$$\hat{T} = \int_A w(x)\hat{S}(x)dx.$$

In other words, given a predicted surface $\hat{S}(\cdot)$, it is reasonable simply to calculate any linear property of this surface and to use the result as the predictor for the corresponding linear property of the true surface $S(\cdot)$. However, this is not the case for prediction of non-linear properties. Note in particular that in practice the point predictor $\hat{S}(\cdot)$ tends to under-estimate peaks and over-estimate troughs in the true surface $S(\cdot)$. Hence, for example, the maximum of $\hat{S}(\cdot)$ would be a poor predictor for the maximum of $S(\cdot)$.

2.4.2 *Extending the Gaussian model*

The Gaussian model discussed so far is, of course, not appropriate for all applications. In later sections, we will discuss a range of non-Gaussian models. Here, we discuss briefly how some of the assumptions may be relaxed whilst remaining within the Gaussian framework.

Firstly, we need to be able to deal with a non-constant mean value surface $\mu(x)$. Technically the simplest case is when $\mu(x)$ is specified by a linear model, $\mu(x) = \sum_{j=1}^{p} \beta_j f_j(x)$, where $f_1(x), \ldots, f_p(x)$ are observed functions of location, x. A special case, known as polynomial trend surface modelling, arises when $f_1(x), \ldots, f_p(x)$ are powers of the spatial coordinates $x_{(1)}$ and $x_{(2)}$. In our opinion, linear or possibly quadratic trend surfaces are occasionally useful as pragmatic descriptions of spatial variation in an overall level of the responses $Y_1, \ldots, Y_n$, but more complicated polynomial trend surfaces are seldom useful, since they often lead to unrealistic extrapolations beyond the convex hull of the sampling design. Another possibility is to define the $f_j(x)'s$ above as functions of observed covariates. Note that this requires covariate measurements also to be available at prediction locations. The procedure of obtaining predictions using a polynomial trend of the coordinates is often called *universal kriging* in the geostatistics literature, while the case when other covariates are used is called *kriging with a trend model* (Goovaerts 1997). Non-linear models for $\mu(x)$ will often be more realistic on physical grounds. However, fitting non-linear models is technically less straightforward than in the linear case and needs to be approached with caution.

Secondly, in some applications we may find empirical evidence of directional effects in the covariance structure. The simplest way to deal with this is by introducing a *geometric anisotropy* into the assumed covariance structure. Physically, this corresponds to a rotation and stretching of the original spatial coordinates. Algebraically, it adds to the model two more parameters: the *anisotropy angle* ψ_A and the *anisotropy ratio* $\psi_R > 1$. These define a transformation of the space of locations $x = (x_{(1)}, x_{(2)})$ according to

$$(x'_{(1)}, x'_{(2)}) = (x_{(1)}, x_{(2)}) \begin{pmatrix} \cos(\psi_A) & -\sin(\psi_A) \\ \sin(\psi_A) & \cos(\psi_A) \end{pmatrix} \begin{pmatrix} 1 & 0 \\ 0 & \psi_R^{-1} \end{pmatrix}$$

and the correlation between two locations is modelled as a function of distance in this transformed space.

A third possible extension is to assume an additional component for the variance, the so-called micro-scale variation, hence in the stationary case with no covariates the model is extended to

$$Y_i = S(x_i) + S_0(x_i) + Z_i : i = 1, ..., n$$

where $S(\cdot)$ and Z_i are as before but additionally $S_0(\cdot)$ is a stationary Gaussian process with rapidly decaying spatial correlation. If we formally assume that $S_0(\cdot)$ is uncorrelated spatial Gaussian white noise (independent of $S(\cdot)$ and the Z_i), then the terms $S_0(x_i)$ and Z_i are indistinguishable. In practice, they will also be indistinguishable if the correlation of $S_0(\cdot)$ decays within a distance smaller than the smallest distance between any two sampling locations. In mining applications the micro-scale component is assumed to be caused by the existence of small nuggets of enriched ore and is approximated by a white noise process. Hence, in practice the term "nugget effect" applied to the independent error term Z_i is interpreted, according to context, as measurement error, micro-scale variation or a non-identifiable combination of the two.

Stationarity itself is a convenient working assumption, which can be relaxed in various ways. A functional relationship between mean and variance can sometimes be resolved by a transformation of the data. When the responses $Y_1, \ldots, Y_n$ are continuous but the Gaussian model is clearly inappropriate, some additional flexibility is obtained by introducing an extra parameter λ defining a Box-Cox transformation of the response. The resulting model assumes that the data, denoted $y = (y_1, ..., y_n)$, can be transformed by

$$\tilde{y}_i = h_\lambda(y_i) = \begin{cases} (y_i^\lambda - 1)/\lambda & \text{if } \lambda \neq 0 \\ \log y_i & \text{if } \lambda = 0, \end{cases} \tag{2.3}$$

such that $(\tilde{y}_1, \ldots, \tilde{y}_n)$ is a realisation from a Gaussian model. De Oliveira, Kedem & Short (1997) propose formal Bayesian methods of inference within this model class, one consequence of which is that their predictions are averages over a range of models corresponding to different values of λ. An alternative approach is to estimate λ, but then hold λ fixed when performing prediction (Christensen, Diggle & Ribeiro Jr 2001). This avoids the difficulty of placing a physical interpretation on a predictive distribution which is averaged over different scales of measurement.

Intrinsic variation, a weaker hypothesis than stationarity, states that the process has stationary increments. This represents a spatial analogue of the random walk model for time series, and is widely used as a default model for discrete spatial variation, see Chapter 3 and Besag, York & Mollié (1991).

Finally, *spatial deformation* methods (Sampson & Guttorp 1992) seek to achieve stationarity by a non-linear transformation of the geographical space, $x = (x_{(1)}, x_{(2)})$.

It is important to remember that the increased flexibility of less restrictive modelling assumptions is bought at a price. In particular, over-complex models fitted to sparse data can easily lead to poor identifiability of model parameters, and to poorer predictive performance than simpler models.

2.5 Parametric estimation of covariance structure

2.5.1 *Variogram analysis*

In classical geostatistics, the standard summary of the second-moment structure of a spatial stochastic process is its variogram. The *variogram* of a stochastic process $Y(\cdot)$ is the function

$$V(x, x') = \frac{1}{2}\text{Var}\{Y(x) - Y(x')\}.$$

For the linear Gaussian model, with $u = ||x - x'||$,

$$V(u) = \tau^2 + \sigma^2\{1 - \rho(u)\}.$$

The basic structural covariance parameters of the linear Gaussian model are the *nugget variance*, τ^2, the *total sill*, $\tau^2 + \sigma^2 = \text{Var}\{Y(x)\}$, and the *range*, ϕ, such $\rho(u) = \rho_0(u/\phi)$. Thus, any reasonable version of the linear Gaussian model will involve at least three covariance parameters. However, we would need abundant data (or contextual knowledge) to justify estimating more

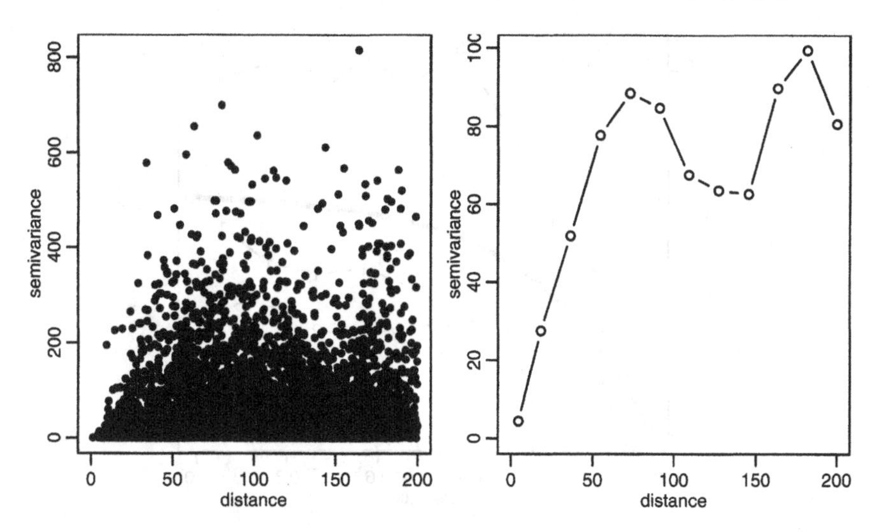

FIGURE 2.9. The variogram cloud (left panel) and binned variogram (right panel) for the Swiss rainfall data.

than three parameters. Note in particular that the Matérn family uses a fourth parameter to determine the differentiability of $S(\cdot)$. Our view is that it is sensible to choose κ from amongst a small set of values to reflect contextual knowledge about the smoothness of $S(\cdot)$, rather than formally to estimate it from sparse data.

The *variogram cloud* of a set of geostatistical data is a scatterplot of the points (u_{ij}, v_{ij}), derived from the quantities

$$\begin{aligned} u_{ij} &= ||x_i - x_j|| \\ v_{ij} &= (y_i - y_j)^2/2. \end{aligned}$$

The left-hand panel of Figure 2.9 shows an example of a variogram cloud, calculated from the Swiss rainfall data. Its diffuse appearance is entirely typical. Note in particular that under the linear Gaussian model, $v_{ij} \sim V(u_{ij})\chi_1^2$ and different v_{ij}'s are correlated. The variogram cloud is therefore unstable, both pointwise and in its overall shape.

When the underlying process has a spatially varying mean $\mu(x)$ the variogram cloud as defined above is not a sensible summary. Instead, we replace the data y_i in the expression for v_{ij} by residuals $r_i = y_i - \hat{\mu}(x_i)$, where $\hat{\mu}(\cdot)$ is an estimate of the underlying mean value surface, typically an ordinary least squares estimate within an assumed linear model.

A more stable variant of the variogram cloud is the *empirical variogram* $\bar{V}(\cdot)$, as illustrated on the right-hand panel of Figure 2.9. For a separation distance u, $\bar{V}(\cdot)$ is obtained by averaging those v_{ij}'s for which $|u - u_{ij}| < h/2$, where h is a chosen bin width. The averaging addresses the first objection to the variogram cloud, namely its pointwise instability, but the difficulties caused by the inherent correlation amongst differ-

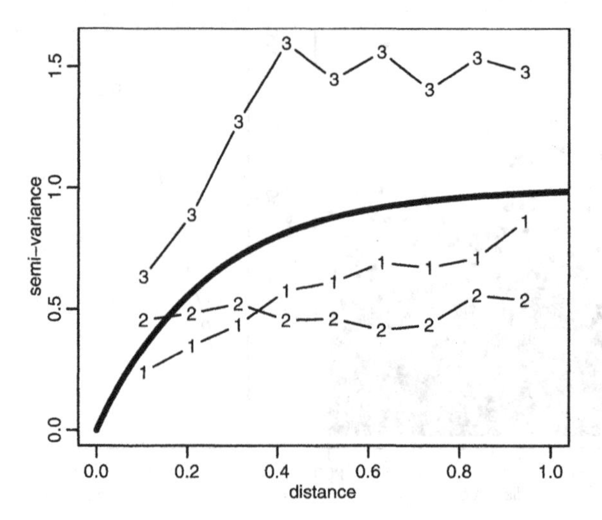

FIGURE 2.10. Empirical variograms from three independent realisations of the same stationary Gaussian process.

ent variogram ordinates remain. Note also that the empirical variogram is necessarily sensitive to mis-specification of the mean value surface $\mu(x)$. Specifically, failure to adjust for long-range variation in the mean response will induce spurious evidence of long-range correlation in $Y(\cdot)$.

Fitting a parametric covariance function to an empirical variogram provides one possible way to estimate covariance parameters. Frequently in practice this is done "by eye", without a formal criterion. Alternatively, ordinary or weighted least squares methods for curve fitting are sometimes used. These methods estimate the covariance parameters θ by minimising

$$S(\theta) = \sum_k w_k[\bar{V}(u_k) - V(u_k;\theta)]^2$$

where $w_k = 1$ for ordinary least squares, whereas for weighted least squares w_k is the number of pairs of measurements which contribute to $\bar{V}(u_k)$. The resulting fits are often visually convincing, but this begs the question of whether matching theoretical and empirical variograms is optimal in any sense. In fact, empirical variograms calculated from typical sizes of data-set are somewhat unstable. To illustrate this, Figure 2.10 compares the empirical variograms from three independent simulations of the same model with the true underlying variogram, where the correlation function is exponential, the parameters $\sigma^2 = 1$, $\phi = 0.25$, $\tau^2 = 0$, and 100 locations randomly distributed in a unit square. The inherently high autocorrelations amongst $\hat{V}(u)$ for successive values of u impart a misleading smoothness into the empirical variograms, suggesting greater precision than is in fact the case.

Parameter estimation via the variogram is a deeply rooted part of classical geostatistical methodology but its popularity is, in our view, misplaced. It does have a role to play in exploratory analysis, at model formulation stage and as a graphical diagnostic. For formal inference, we prefer likelihood-based methods. These have the compelling (to us) advantage that they are optimal under the stated assumptions, although they are computationally expensive for large data-sets, and a legitimate concern is that they may lack robustness. The likelihood function also plays a central role in Bayesian inference, in which estimation and prediction are naturally combined. We discuss this in greater detail in Section 2.7.

2.5.2 Maximum likelihood estimation

Under the Gaussian model

$$Y \sim N(F\beta, \sigma^2 R + \tau^2 I)$$

where F is the $n \times p$ matrix of covariates, β is the vector of parameters, and R depends on (ϕ, κ). The log-likelihood function is

$$\begin{aligned} l(\beta, \tau^2, \sigma^2, \phi, \kappa) \;\; \propto \;\; & -0.5\{\log|(\sigma^2 R + \tau^2 I)| \\ & +(y - F\beta)^{\mathrm{T}}(\sigma^2 R + \tau^2 I)^{-1}(y - F\beta)\}, \end{aligned} \qquad (2.4)$$

maximisation of which yields the maximum likelihood estimates of the model parameters.

Computational details are as follows. Firstly, we reparameterise to $\nu^2 = \tau^2/\sigma^2$ and denote $V = (R + \nu^2 I)$. Given V, the log-likelihood function is maximised for

$$\hat{\beta}(V) = (F^{\mathrm{T}}V^{-1}F)^{-1}F^{\mathrm{T}}V^{-1}y$$

and

$$\hat{\sigma}^2(V) = n^{-1}(y - F\hat{\beta})^{\mathrm{T}}V^{-1}(y - F\hat{\beta}).$$

Hence, substituting $(\hat{\beta}(V), \hat{\sigma}^2(V))$ into the log-likelihood function, we obtain the reduced log-likelihood

$$l(\nu^2, \phi, \kappa) \propto -0.5\{n \log|\hat{\sigma}^2(V)| + \log|V|\}.$$

This must then be optimised numerically, followed by back-substitution to obtain $\hat{\sigma}^2$ and $\hat{\beta}$. In practice, for the Matérn correlation function we suggest choosing κ from the discrete set $\{0.5, 1, 1.5, 2, 2.5, ..., K/2\}$ for some small integer K.

If geometric anisotropy parametrised by (ψ_A, ψ_R) is included in the model, the same procedure is used, except that the additional parameters need to be incorporated into the matrix R, thereby adding two dimensions to the numerical maximisation of the likelihood.

For the transformed Gaussian model defined by (2.3), the associated log-likelihood is

$$\begin{aligned}\ell(\beta, \sigma^2, \phi, \nu^2, \kappa, \lambda) &= (\lambda - 1)\sum_{i=1}^{n} \log y_i - 0.5 \log |\sigma^2 V| \\ &\quad -0.5(h_\lambda(y) - F\beta)^{\mathrm{T}}\{\sigma^2 V\}^{-1}(h_\lambda(y) - F\beta)\}.\end{aligned}$$

Here we use the procedure above, but adding optimisation with respect to λ in the numerical maximisation.

A popular variant of maximum likelihood estimation is *restricted maximum likelihood estimation* (REML). Under the assumed model for $\mathrm{E}[Y] = F\beta$, we can transform the data linearly to $Y^* = AY$ such that the distribution of Y^* does not depend on β. Then, the REML principle is to estimate $\theta = (\nu^2, \sigma^2, \phi, \kappa)$ by maximum likelihood applied to the transformed data Y^*. We can always find a suitable matrix A without knowing the true values of β or θ, for example a projection to ordinary least squares residuals,

$$A = I - F(F^{\mathrm{T}}F)^{-1}F^{\mathrm{T}}.$$

The REML estimators for θ is computed by maximising

$$\begin{aligned}l^*(\theta) \;\propto\; & -0.5\{\log|\sigma^2 V| - \log|F^{\mathrm{T}}\{\sigma^2 V\}^{-1}F| \\ & +(y - F\tilde{\beta})^T\{\sigma^2 V\}^{-1}(y - F\tilde{\beta}))\},\end{aligned}$$

where $\tilde{\beta} = \hat{\beta}(V)$. Note the extra determinant term by comparison with the ordinary log-likelihood given by (2.4).

REML was introduced in the context of variance components estimation in designed experiments (Patterson & Thompson 1971) and some early references in the geostatistical context are Kitanidis (1983) and Zimmerman (1989). In general, it leads to less biased estimators of variance parameters in small samples (for example, the elementary unbiased sample variance is a REML estimator). Note that $l^*(\theta)$ depends on F, and therefore on a correct specification of the model for $\mu(x)$. For designed experiments, the specification of the mean $\mu(x)$ is usually not problematic. However, in the spatial setting the specification of the mean $\mu(x)$ is often a pragmatic choice. Although REML is widely recommended for geostatistical models, our experience has been that it is more sensitive than ML to misspecification of the model for $\mu(x)$.

Another generic likelihood-based idea which is useful in the geostatistical setting is that of profile likelihoods. In principle, variability of parameter estimators can be investigated by inspection of the log-likelihood surface. However, the typical dimension of this surface does not allow direct inspection. Suppose, in general, that we have a model with parameters (α, ψ) and denote its likelihood by $L(\alpha, \psi)$. To inspect the likelihood for α, we replace the nuisance parameters ψ by their ML estimators $\hat{\psi}(\alpha)$, for each value of

α. This gives the *profile likelihood* for α,

$$L_p(\alpha) = L(\alpha, \hat{\psi}(\alpha)) = \max_{\psi}(L(\alpha, \psi)).$$

The profile log-likelihood can be used to calculate approximate confidence intervals for individual parameters, exactly as in the case of the ordinary log-likelihood for a single parameter model.

2.6 Plug-in prediction

We use this term to mean the simple approach to prediction whereby estimates of unknown model parameters are plugged into the prediction equations as if they were the truth. This tends to be optimistic in the sense that it leads to an under-estimation of prediction uncertainty by ignoring variability between parameter estimates and their true, unknown values. Nevertheless, it is widely used, corresponds to standard geostatistical methods collectively known as kriging, and is defensible in situations where varying model parameters over reasonable ranges produces only small changes in the sizes of the associated prediction variances.

2.6.1 The Gaussian model

For the Gaussian model we have seen that the minimum MSE predictor for $T = S(x_0)$ is

$$\hat{T} = \mu + \sigma^2 \mathbf{r}^{\mathrm{T}}(\tau^2 I + \sigma^2 R)^{-1}(y - \mu \mathbf{1})$$

with prediction variance

$$\mathrm{Var}[T|y] = \sigma^2 - \sigma^2 \mathbf{r}^{\mathrm{T}}(\tau^2 I + \sigma^2 R)^{-1}\sigma^2 \mathbf{r}.$$

A *plug-in* prediction consists of replacing the true parameters in the prediction equations above by their estimates. As noted earlier, simple kriging is prediction where estimates of the mean and covariance parameters are plugged-in. Another approach often used in practice is *ordinary kriging*, which only requires covariance parameters to be plugged-in (Journel & Huijbregts 1978). Ordinary kriging uses a linear predictor which minimises the mean square prediction error under an unbiasedness constraint which implies that the prediction weights must sum to one. This filters out the mean parameter from the expression for the predictor.

2.6.2 The transformed Gaussian model

For the Box-Cox transformed Gaussian model, assume $Y(x_0)$ is the target for prediction, and denote $T_\lambda = h_\lambda(Y(x_0))$. The minimum mean square error predictor $\hat{T}_\lambda$ and the corresponding prediction variance $\text{Var}[T_\lambda \mid y]$ are found as above using simple kriging. Back-transforming to the original scale is done using formulas for moments. For $\lambda = 0$ we use properties of the exponential of a normal distribution and get

$$\hat{T} = \exp(\hat{T}_0 + 0.5\text{Var}[T_0 \mid y])$$

with prediction variance

$$\text{Var}[T|y] = \exp(2\hat{T}_0 + \text{Var}[T_0 \mid y])(\exp(\text{Var}[T_0 \mid y]) - 1).$$

For $\lambda > 0$ we can approximate $\hat{T}$ and $\text{Var}[T|y]$ by a sum of moments for the normal distribution. For $\lambda = 0.5$ we get

$$\hat{T} \approx (0.5\hat{T}_{0.5} + 1)^2 + 0.25\text{Var}[T_{0.5} \mid y]$$

with prediction variance

$$\text{Var}[T|y] \approx (0.5\hat{T}_{0.5}+1)^4+1.5(0.5\hat{T}_{0.5}+1)^2\text{Var}[T_{0.5} \mid y]+3(\text{Var}[T_{0.5} \mid y])^2/16.$$

Alternatively, back-transformation to the original scale can be done by simulation as discussed in the next sub-section.

2.6.3 Non-linear targets

In our experience, the plug-in approach and the Bayesian approach presented in the next section usually give similar point predictions when predicting $T = S(x_0)$, but often the prediction variances differ and the two approaches can produce very different results when predicting non-linear targets.

Consider prediction of the non-linear target $T = T(S^*)$ where S^* are values of $S(\cdot)$ at some locations of interest (for example, a fine grid over the entire area). A general way to calculate the predictor $\hat{T}$ is by simulation. The procedure consists of the following three steps.

- Calculate $\text{E}[S^*|y]$ and $\text{Var}[S^*|y]$ using simple kriging.
- Simulate $s^*(1), \ldots, s^*(m)$ from $[S^*|y]$ (multivariate Gaussian).
- Approximate the minimum mean square error predictor

$$\text{E}[T(S^*)|y] \approx \frac{1}{m}\sum_{j=1}^{m} T(s^*(j)).$$

For the transformed Gaussian model we use a procedure similar to above, we just need to back-transform the simulations by $h_\lambda^{-1}(\cdot)$ before taking averages.

2.7 Bayesian inference for the linear Gaussian model

Bayesian inference treats parameters in the model as random variables, and therefore makes no formal distinction between parameter estimation problems and prediction problems. This provides a natural means of allowing for parameter uncertainty in predictive inference.

2.7.1 Fixed correlation parameters

To derive Bayesian inference results for the linear Gaussian model, we first consider the situation in which we fix $\tau^2 = 0$, all other parameters in the correlation function have known values, and we allow for uncertainty only in the parameters β and σ^2. In this case the predictive distributions can be derived analytically.

For fixed ϕ, the conjugate prior family for (β, σ^2) is the Gaussian-Scaled-Inverse-χ^2. This specifies priors for β and σ^2 with respective distributions

$$[\beta|\sigma^2, \phi] \sim N\left(m_b, \sigma^2 V_b\right) \quad \text{and} \quad [\sigma^2|\phi] \sim \chi^2_{ScI}\left(n_\sigma, S^2_\sigma\right),$$

where a $\chi^2_{ScI}(n_\sigma, S^2_\sigma)$ distribution has density of the form

$$\pi(z) \propto z^{-(n_\sigma/2+1)} \exp(-n_\sigma S^2_\sigma/(2z)), \quad z > 0.$$

As a convenient shorthand, we write this as

$$[\beta, \sigma^2|\phi] \sim N\chi^2_{ScI}\left(m_b, V_b, n_\sigma, S^2_\sigma\right), \tag{2.5}$$

Using Bayes' Theorem, the prior above is combined with the likelihood given by (2.4) and the resulting posterior distribution of the parameters is:

$$[\beta, \sigma^2|y, \phi] \sim N\chi^2_{ScI}\left(\tilde{\beta}, V_{\tilde{\beta}}, n_\sigma + n, S^2\right), \tag{2.6}$$

where $V_{\tilde{\beta}} = (V_b^{-1} + F^{\mathrm{T}}R^{-1}F)^{-1}$, $\tilde{\beta} = V_{\tilde{\beta}}(V_b^{-1}m_b + F^{\mathrm{T}}R^{-1}y)$ and

$$S^2 = \frac{n_\sigma S^2_\sigma + m_b^{\mathrm{T}} V_b^{-1} m_b + y^{\mathrm{T}} R^{-1} y - \tilde{\beta}^{\mathrm{T}} V_{\tilde{\beta}}^{-1} \tilde{\beta}}{n_\sigma + n}. \tag{2.7}$$

The predictive distribution of the signal at an arbitrary set of locations, say $S^* = (S(x_{n+1}), \ldots, S(x_{n+q}))$, is obtained by integration,

$$p(s^*|y, \phi) \quad = \quad \int\int p(s^*|y, \beta, \sigma^2, \phi)\, p(\beta, \sigma^2|y, \phi)\, d\beta d\sigma^2,$$

where $[s^*|y, \beta, \sigma^2, \phi]$ is multivariate Gaussian with mean and variance given by (2.1) and (2.2) respectively. The integral above yields a q-dimensional multivariate-t distribution defined by:

$$\begin{aligned}
[S^*|y,\phi] &\sim t_{n_\sigma+n}\left(\mu^*, S^2\Sigma^*\right), \\
\mathrm{E}[S^*|y,\phi] &= \mu^*, \\
\mathrm{Var}[S^*|y,\phi] &= \frac{n_\sigma+n}{n_\sigma+n-2}\, S^2\Sigma^*,
\end{aligned} \tag{2.8}$$

where S^2 is given by (2.7) and μ^* and Σ^* are

$$\begin{aligned}
\mu^* &= (F_0 - r^{\mathrm{T}}R^{-1}F)V_{\tilde{\beta}}V_b^{-1}m_b \\
&\quad + \left[r^{\mathrm{T}}R^{-1} + (F_0 - r^{\mathrm{T}}R^{-1}F)V_{\tilde{\beta}}F^{\mathrm{T}}R^{-1}\right] y, \\
\Sigma^* &= R_0 - r^{\mathrm{T}}R^{-1}r + (F_0 - r^{\mathrm{T}}R^{-1}F)(V_b^{-1} + V_{\hat{\beta}}^{-1})^{-1}(F_0 - r^{\mathrm{T}}R^{-1}F)^{\mathrm{T}}.
\end{aligned}$$

The three components in the formula for the prediction variance Σ^* can be interpreted as the variability a priori, the reduction due to the conditioning on the data, and the increase due to uncertainty in the value of β, respectively.

It may be difficult to elicit informative priors in practice, and flat or non-informative improper priors might therefore be adopted. A non-informative prior often used in Bayesian analysis of linear models is $\pi(\beta, \sigma^2) \propto 1/\sigma^2$ (see for example, O'Hagan (1994)). Formal substitution of $V_b^{-1} = 0$ and $n_\sigma = 0$ into the formulas above for the posterior and predictive distributions gives the equivalent formulas for the non-informative prior, except that the degrees of freedom in the χ^2 posterior distribution and the multivariate-t predictive distribution are $n-p$ where p is the dimension of β, rather than n.

For the transformed Gaussian model, when $\lambda > 0$, we can back-transform predictions to the original scale using formulas for moments of the t-distribution, similar to the approach in Section 2.6. Note, however, that the exponential of a t-distribution does not have finite moments, hence when $\lambda = 0$ the minimum mean square error predictor does not exist. Prediction of non-linear targets is done using a procedure similar to the one in Section 2.6.3.

2.7.2 Uncertainty in the correlation parameters

More realistically, we now allow for uncertainty in all of the model parameters. We first consider the case of a model without measurement error, i.e. $\tau^2 = 0$ and a single correlation parameter ϕ. We adopt a prior $\pi(\beta, \sigma^2, \phi) = \pi(\beta, \sigma^2|\phi)\ \pi(\phi)$, the product of (2.5) and a proper density for ϕ. In principle a continuous prior $\pi(\phi)$ would be assigned. However, in practice we always use a discrete prior, obtained by discretising the distribution of ϕ in equal width intervals. The posterior distribution for the parameters is then given by

$$p(\beta, \sigma^2, \phi|y) = p(\beta, \sigma^2|y, \phi)\, p(\phi|y)$$

with $[\beta, \sigma^2|y, \phi]$ given by (2.6) and

$$p(\phi|y) \propto \pi(\phi)\, |V_{\tilde{\beta}}|^{\frac{1}{2}}\, |R|^{-\frac{1}{2}}\, (S^2)^{-\frac{n+n_\sigma}{2}}, \tag{2.9}$$

where $V_{\tilde{\beta}}$ and S^2 are given by (2.6) and (2.7) respectively. For the case where the prior is $\pi(\beta, \sigma^2, \phi) \propto \pi(\phi)/\sigma^2$, the equation above holds with $n_\sigma = -p$. Berger, De Oliveira & Sansó (2001) use a special case of this as a non-informative prior for the parameters of a spatial Gaussian process

To simulate samples from this posterior, we proceed as follows. We apply (2.9) to compute posterior probabilities $p(\phi|y)$ noting that in practice the support set will be discrete. We then simulate a value of ϕ from $[\phi|y]$, attach the sampled value to $[\beta, \sigma^2|y, \phi]$ and obtain a simulation from this distribution. By repeating the simulation as many times as required, we obtain a sample of triplets (β, σ^2, ϕ) from the joint posterior distribution of the model parameters.

The predictive distribution for the value, $S_0 = S(x_0)$ say, of the signal process at an arbitrary location x_0 is given by

$$\begin{aligned} p(s_0|y) &= \iiint p(s_0, \beta, \sigma^2, \phi|y)\, d\beta\, d\sigma^2\, d\phi \\ &= \iiint p\big(s_0, \beta, \sigma^2|y, \phi\big)\; d\beta\, d\sigma^2\, p(\phi|y)\; d\phi \\ &= \int p(s_0|y, \phi)\, p(\phi|y)\, d\phi. \end{aligned}$$

The discrete prior for ϕ allows analytic calculation of the moments of this predictive distribution. For each value of ϕ we compute the moments of the multivariate-t distribution given by (2.8) and calculate their weighted sum with weights given by the probabilities $p(\phi|y)$.

To sample from this predictive distribution, we proceed as follows. We compute the posterior probabilities $p(\phi|y)$ on the discrete support set for $[\phi]$, and simulate values of ϕ from $[\phi|y]$. Attaching a sampled value of ϕ to $[S_0|y, \phi]$ and simulating from this distribution we obtain a realisation from the predictive distribution.

Finally, when $\tau^2 > 0$, in practice we use a discrete joint prior $[\phi, \nu^2]$, where $\nu^2 = \tau^2/\sigma^2$. This adds to the computational load, but introduces no new principles. Similarly, if we wish to incorporate additional parameters in the covariance structure of $S(\cdot)$, we would again use a discretisation method to render the computations feasible.

In principle, the prior distributions for the parameters should reflect scientific prior knowledge. In practice, we will often be using the Bayesian framework pragmatically, under a vague prior specification. However, a

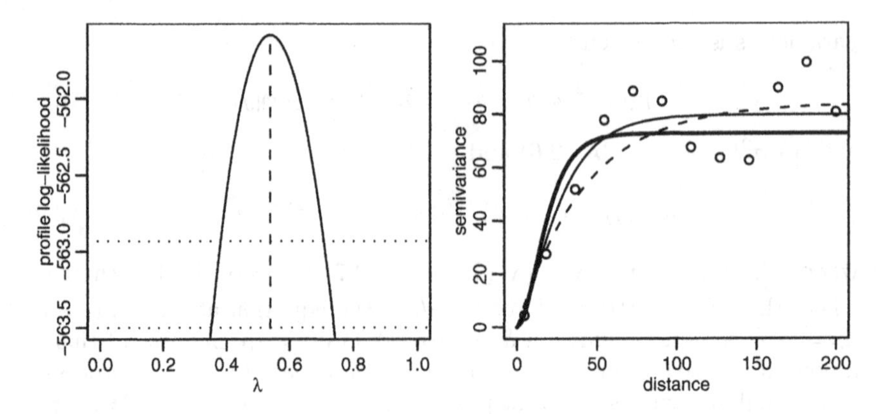

FIGURE 2.11. Left panel: profile likelihood for transformation parameter λ for the model with Matérn correlation function with $\kappa = 1$. Right panel: Estimated variograms for transformed ($\lambda = 0.5$) data (open circles), compared with the theoretical Matérn model with parameters equal to the maximum likelihood estimates. The three fits correspond to $\kappa = 0.5$ (dashed line), $\kappa = 1$ (thick solid line), $\kappa = 2$ (thin solid line).

word of caution is necessary here, as we have found that even apparently vague prior specifications can materially affect the corresponding posteriors. It seems to be a general feature of geostatistical problems that the models are poorly identified, in the sense that widely different combinations of parameter values lead to very similar fits. This may not matter if parameter estimates themselves, as opposed to the prediction target T, are not of direct interest. Also, the Bayesian paradigm at least brings this difficulty into the open, whereas plugging in more or less arbitrary point estimates merely hides the problem.

2.8 A Case Study: the Swiss rainfall data

In this case study, we follow convention by using only the first 100 of the data-locations in the Swiss rainfall data for model formulation. We consider a transformed Gaussian model, with a Matérn correlation structure.

Table 2.1 shows the maximum likelihood estimates of the Box-Cox transformation parameter λ, holding the Matérn shape parameter κ fixed at each of the three values $\kappa = 0.5, 1, 2$. The consistent message is that $\lambda = 0.5$, or a square root transformation, is a reasonable choice. The profile log-likelihood for λ shown in the left-hand panel of Figure 2.11 indicates that neither the log-transformation $\lambda = 0$, nor an untransformed Gaussian assumption ($\lambda = 1$) is tenable for these data. The right-hand panel of Figure 2.11 shows the empirical and fitted variograms, for each of $\kappa = 0.5, 1, 2$. Visually, there is little to choose amongst the three fits.

κ	$\hat{\lambda}$	$\log \hat{L}$
0.5	0.496	-564.857
1	0.540	-561.579
2	0.561	-563.115

TABLE 2.1. Maximum likelihood estimates $\hat{\lambda}$ and the corresponding values of the log-likelihood function $\log \hat{L}$ for the Swiss rainfall data, assuming different values of the Matérn shape parameter κ.

κ	$\hat{\beta}$	$\hat{\sigma}^2$	$\hat{\phi}$	$\hat{\tau}^2$	$\log \hat{L}$
0.5	21.205	83.865	42.388	0	-564.858
1.0	22.426	79.694	17.583	0	-561.664
2.0	23.099	72.698	8.358	0	-563.292

TABLE 2.2. Maximum likelihood estimates $\hat{\beta}$, $\hat{\phi}$, $\hat{\sigma}$, $\hat{\tau}^2$ and the corresponding value of the likelihood function $\log \hat{L}$ for the Swiss rainfall data, assuming different values of the Matérn parameter κ, and transformation parameter $\lambda = 0.5$.

Table 2.2 shows maximum likelihood estimates for the model with $\lambda = 0.5$. The overall conclusion is that $\kappa = 1$ gives a better fit than $\kappa = 0.5$ and $\kappa = 2$. Furthermore, in each case $\hat{\tau}^2 = 0$. Figure 2.12 shows the profile log-likelihoods of the two covariance parameters σ^2, ϕ holding κ, λ and τ^2 fixed at these values. Note in particular the wide, and asymmetric, confidence intervals for the signal variance σ^2 and the range parameter ϕ. These serve to warn against over-interpretation of the corresponding point estimates.

Figure 2.13 maps the point predictions of rainfall values and associated

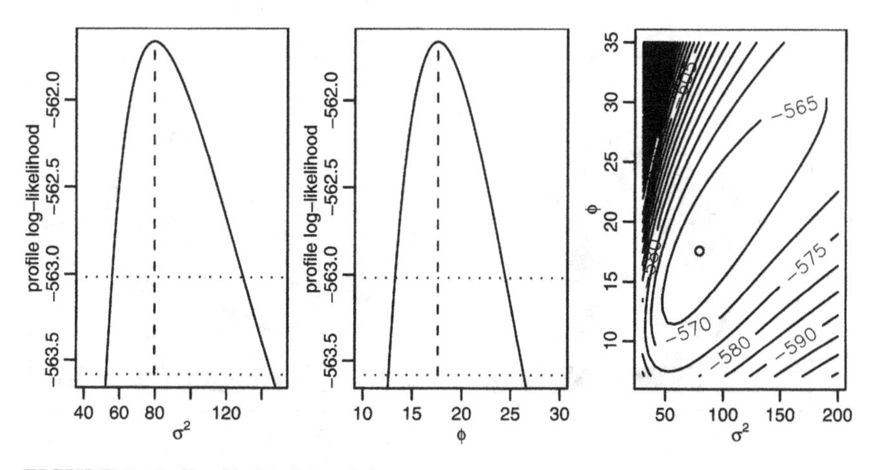

FIGURE 2.12. Profile likelihood for covariance parameters in the Matérn model fitted to the Swiss rainfall data with $\kappa = 1$ and $\lambda = 0.5$. Left panel σ^2, middle panel ϕ, right panel the 2-D profile likelihood.

prediction variances from a plug-in prediction using the transformed Gaussian model with $\lambda = 0.5$ and $\kappa = 1$. The grid spacing for prediction corresponds to a distance of 5 km between adjacent prediction locations. The values of the prediction variances shows a positive association with predicted values, as a consequence of the transformation adopted; recall that in the untransformed Gaussian model, the prediction variance depends only on the model parameters and the study design, and not directly on the measured values.

We now turn to the Bayesian analysis, adopting the prior $\pi(\beta, \sigma^2|\phi) \propto 1/\sigma^2$ and a discrete uniform prior for ϕ with 101 points equally spaced in the interval $[0; 100]$. The posterior distribution for ϕ is then obtained by computing (2.9) for each discrete value, and standardising such that the probabilities add to one. The left-hand panel of Figure 2.14 shows the uniform prior adopted and the posterior distribution obtained for this data-set. The right-hand panel of Figure 2.14 displays variograms based on different summaries of the posterior $[\sigma^2, \phi|y]$ and on the ML estimates $(\hat{\sigma}^2, \hat{\phi})$. The differences between the Bayesian estimates reflect the asymmetry in the posterior distributions of ϕ and σ^2. Note that in all three cases the Bayesian estimate of σ^2 is greater than the ML estimate.

Values of the parameters ϕ and σ^2 sampled from the posterior are displayed by the histograms in the left and centre panels of Figure 2.15. The right-hand panel of Figure 2.15 shows that there is a strong correlation in the posterior, despite the fact that priors for these two parameters are independent. This echoes the shape of the two-dimensional profile likelihood shown earlier in Figure 2.12. Similar results were obtained for other choices of prior.

To predict the values of rainfall in a grid of points over Switzerland we

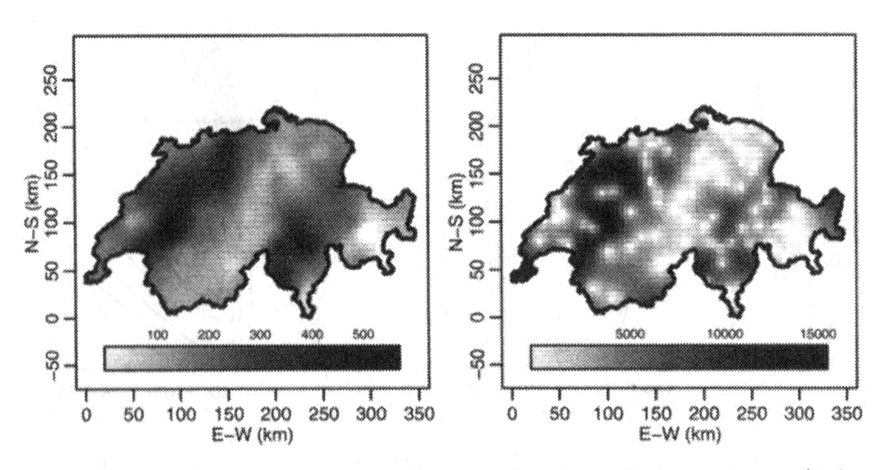

FIGURE 2.13. Maps of predictions (left panel) and prediction variances (right panel) for the Swiss rainfall data.

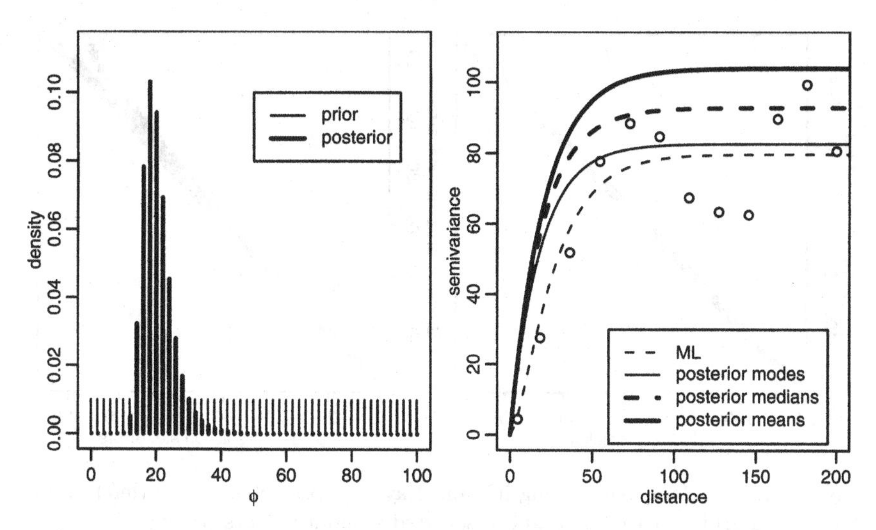

FIGURE 2.14. Left panel: uniform prior and corresponding posterior distribution. Right panel: variograms based on summaries of the posterior and on the ML estimator.

can compute moments analytically, as described in Section 2.7.1 and Section 2.7.2. Figure 2.16 shows a comparison between "plug-in" and Bayesian point predictions (left panel) and their standard errors (right panel). The strong concentration of points along the diagonal in the left-hand panel of Figure 2.16 shows that, for this particular example, the Bayesian point predictions do not differ too much from the "plug-in" predictions. However, as indicated in the right-hand panel of Figure 2.16 there are differences in the estimated uncertainty associated with the predictions, with the plug-in

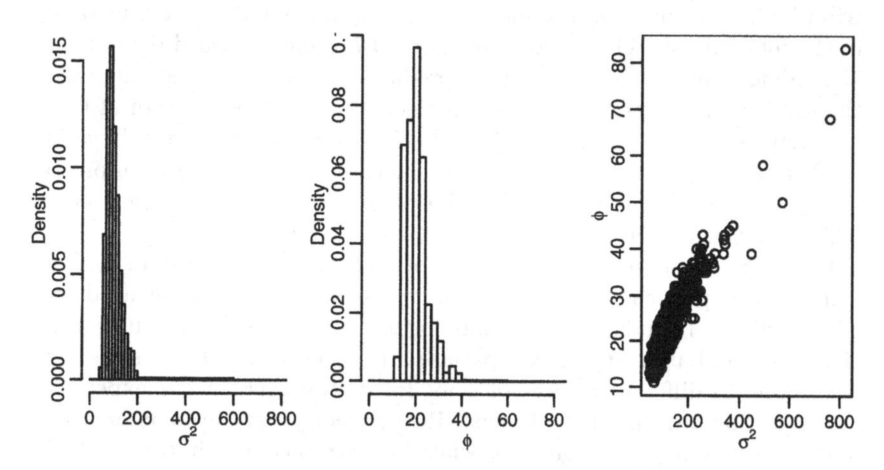

FIGURE 2.15. Histograms for samples from the posteriors for the parameters σ^2 (left) and ϕ (middle), and corresponding scatterplot.

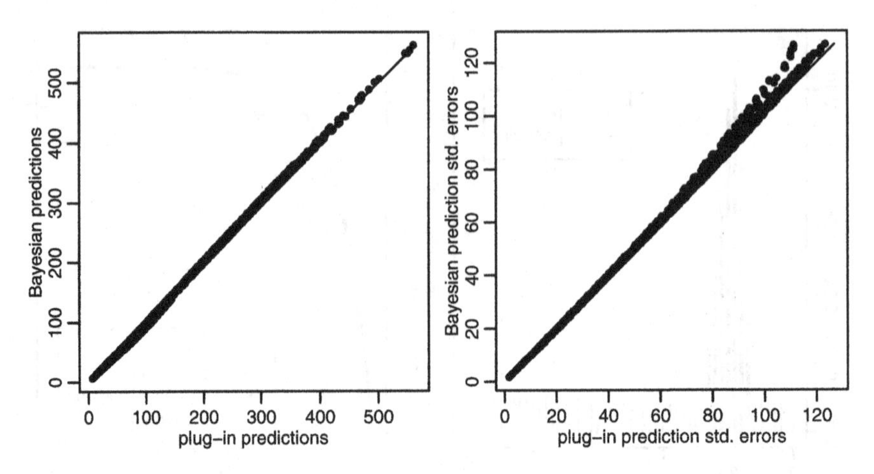

FIGURE 2.16. Comparing "plug-in" and Bayesian predicted values (left) in a 5 km spaced grid over the area, and associated standard errors (right).

variances tending to slightly under-estimate the variance of the predictive distribution, especially where the prediction variance itself is relatively large.

Inferences about non-linear functionals can be performed by sampling from the predictive distribution and processing the sampled values according to the functional of interest. This generates a sample from the posterior distribution of the target for prediction. As an example, consider inference for the target $T_{max} = max\{Y(x) : x \in A\}$, the maximum rainfall over the whole of Switzerland. In practice we redefined T_{max} to be the maximum over the 5 km spaced grid. Taking 2000 simulations from the predictive distribution and computing the maximum for each simulation we find values in the interval $[531, 1114]$ with a mean of 667.4 and standard deviation of 73.9. Simulations from the "plug-in" predictive distribution generated with the same seed for the random number generator showed a mean of 655.8 and standard deviation of 67.4. So for this prediction target the Bayesian prediction is larger than the plug-in prediction. Also, the Bayesian prediction standard error is larger than the plug-in prediction standard error, which is often seen in practice, but is not always the case.

From our experience with a variety of real and simulated data-sets, we consider this particular data-set to be an exceptionally well behaved one. The profile likelihoods are sharp and not too wide. No extra residual variation was found after fitting the spatial part of the model. The results were insensitive to different choices of prior for ϕ. However, in our experience this situation is somewhat atypical. Rather, noisy data are common and inferences tend to have greater associated uncertainty than in this example. In these situations, the discrepancy between Bayesian and plug-in methods becomes more pronounced.

In the Bayesian analysis reported here we have used vague priors. Ideally, more informative priors relevant to the problem at hand should be considered, although elicitation of such priors is often a difficult task.

2.9 Generalised linear spatial models

The classical generalised linear model (GLM) is defined for a set of mutually independent responses $Y_1, ..., Y_n$. The expectations $\mu_i = \mathrm{E}[Y_i]$ are specified by a *linear predictor* $h(\mu_i) = \sum_{j=1}^{k} f_{ij}\beta_j$, in which $h(\cdot)$ is a known function, called the *link* function (McCullagh & Nelder 1989). An important extension of this basic class of models is the *generalised linear mixed model* or GLMM (Breslow & Clayton 1993), in which $Y_1, \ldots, Y_n$ are mutually independent conditional on the realised values of a set of latent random variables $U_1, \ldots, U_n$, and the conditional expectations are given by $h(\mu_i) = U_i + \sum_{j=1}^{k} f_{ij}\beta_j$. A *generalised linear spatial model* (GLSM) is a GLMM in which the $U_1, \ldots, U_n$ are derived from a spatial process $S(\cdot)$. This leads to the following model-specification.

Let $S(\cdot) = \{S(x) : x \in A\}$ be a Gaussian stochastic process with $\mathrm{E}[S(x)] = \sum_{j=1}^{p} f_j(x)\beta_j$, $\mathrm{Var}[S(x)] = \sigma^2$ and $\rho(u) = \mathrm{Corr}[S(x), S(x')]$ where $u = \|x - x'\|$. Assume that measurements $Y_1, \ldots, Y_n$ are conditionally independent given $S(\cdot)$, with conditional expectations μ_i and $h(\mu_i) = S(x_i), i = 1, \ldots, n$, for a known link function $h(\cdot)$. In this model the signal process is $\{h^{-1}(S(x)) : x \in A\}$.

As in the case of the classical GLM, the GLSM embraces the linear Gaussian model as a special case, whilst providing a natural extension to deal with response variables for which a standard distribution other than the Gaussian more accurately describes the sampling mechanism involved. In what follows, we focus on the Poisson-log-linear model for count data and the logistic model for binomial data.

We denote the regression parameters by β and covariance parameters in the model by θ. We write $Y = (Y_1, ..., Y_n)^{\mathrm{T}}$ for the observed responses at locations $x_1, \ldots, x_n$ in the sampling design, $S = (S(x_1), ..., S(x_n))^{\mathrm{T}}$ for the unobserved values of the underlying process at $x_1, \ldots, x_n$, and S^* for the values of $S(\cdot)$ at all other locations of interest, typically a fine grid of locations covering the study region. The conditional independence structure of the GLSM is then indicated by the graph below. The likelihood for a model of this kind is in general not expressible in closed form, but only as a high-dimensional integral

$$L(\beta, \theta) = \int \prod_{i=1}^{n} g(y_i; h^{-1}(s_i)) p(s; \beta, \theta) ds_1, \ldots, s_n, \tag{2.10}$$

where $g(y; \mu)$ denotes the density of the error distribution parameterised by the mean μ, and $p(s; \beta, \theta)$ is the multivariate Gaussian density for the vector

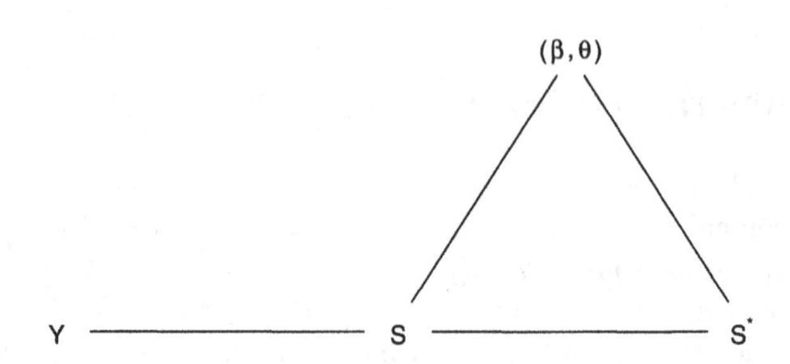

S. The integral above is also the normalising constant in the conditional distribution of $[S|y, \beta, \theta]$,

$$p(s \mid y, \beta, \theta) \propto \prod_{i=1}^{n} g(y_i; h^{-1}(s_i))p(s; \beta, \theta). \tag{2.11}$$

In practice, the high dimensionality of the integral prevents direct calculation of the predictive distribution $[S^* \mid y, \beta, \theta]$.

Standard methods of approximating the integral (2.10) and hence evaluating (2.11) are of unknown accuracy in the geostatistical setting, but Markov chain Monte Carlo methods (see Chapter 1) provide a possible solution.

2.9.1 Prediction in a GLSM

Assume first that the parameters in the model are known. From the figure with the graphical model above we see that prediction of $T = T(S^*)$ can be separated into three steps.

- Simulate $s(1), \ldots, s(m)$ from $[S|y]$ (using MCMC).
- Simulate $s^*(j)$ from $[S^*|s(j)]$, $j = 1, \ldots, m$ (multivariate Gaussian).
- Approximate the minimum mean square error predictor

$$\mathrm{E}[T(S^*)|y] \approx \frac{1}{m}\sum_{j=1}^{m} T(s^*(j)).$$

Whenever possible, it is desirable to replace Monte Carlo sampling by direct evaluation. For example, if it is possible to calculate $\mathrm{E}[T(S^*)|s(j)]$, $j =$

$1, \ldots, m$ directly, we would use the approximation

$$\mathrm{E}[T(S^*)|y] \approx \frac{1}{m} \sum_{j=1}^{m} \mathrm{E}[T(S^*)|s(j)],$$

thereby reducing the Monte Carlo error due to simulation.

To simulate from $[S \mid y]$ we use the *truncated Langevin-Hastings algorithm* as in Christensen, Møller & Waagepetersen (2001); see also Sections 1.3.1 and 4.6.2. This algorithm uses gradient information in the proposal distribution and has been found to work well in practice by comparison with a random walk Metropolis algorithm. First we make a *reparameterisation* defining $S = F^{\mathrm{T}}\beta + \Omega^{1/2}\Gamma$ where $\Omega^{1/2}$ is a square root of $\Omega = \mathrm{Var}[S]$, say a Cholesky factorisation, and a priori $\Gamma \sim N(0, I)$. Using an MCMC-algorithm to obtain a sample $\gamma_1, \ldots, \gamma_n$ from $[\Gamma \mid y]$, we multiply by $\Omega^{1/2}$ and obtain a sample $s(1), \ldots, s(m)$ from $[S|y]$.

The MCMC-algorithm used is a Metropolis-Hastings algorithm where all components of Γ are updated simultaneously. The proposal distribution is a multivariate Gaussian distribution with mean $m(\gamma) = \gamma + (\delta/2)\nabla(\gamma)$ where $\nabla(\gamma) = \frac{\partial}{\partial\gamma} \log f(\gamma \mid y)$, and variance δI_n. For a GLSM with canonical link function h, the gradient $\nabla(\gamma)$ has the following simple form:

$$\nabla(\gamma) = \frac{\partial}{\partial\gamma} \log f(\gamma \mid y) = -\gamma + (\Omega^{1/2})^{\mathrm{T}}\{y - h^{-1}(s)\}, \tag{2.12}$$

where $s = F^{\mathrm{T}}\beta + \Omega^{1/2}\gamma$ and h^{-1} is applied coordinatewise. If we modify the gradient $\nabla(\gamma)$ (by truncating, say) such that the term $\{y - h^{-1}(s)\}$ is bounded, the algorithm can be shown to be geometrically ergodic, and a Central Limit Theorem therefore exits. The Central Limit Theorem with asymptotic variance estimated by Geyer's monotone sequence estimate (Geyer 1992), can be used to assess the Monte Carlo error of the calculated prediction, cf. Section 1.5.3. This algorithm is not specific to the canonical case since the formula in (2.12) can be generalised to accommodate models with a non-canonical link function.

In practice one has to choose the proposal variance δ. We tune the algorithm by running a few test runs and choosing δ such that approximately 60% of the proposals are accepted. To avoid storing a large number of high-dimensional simulations $s(1), \ldots, s(m)$ we also thin the sample such that, say, only every 100th simulation is stored.

2.9.2 Bayesian inference for a GLSM

First we consider Bayesian inference for a GLSM, using the Gaussian-Scaled-Inverse-χ^2 prior for (β, σ^2) defined in (2.5), holding ϕ fixed. The marginal density of S, obtained by integrating over β and σ^2, becomes an n-dimensional multivariate-t density, $t_{n_\sigma}(m_b, S_\sigma^2(R + FV_bF^{\mathrm{T}}))$. Therefore the posterior density of S is

$$p(s \mid y) \propto \prod_{i=1}^{n} g(y_i; h^{-1}(s_i))p(s) \tag{2.13}$$

where $p(s)$ is the marginal density of S.

In order to obtain a sample $s(1), \ldots, s(m)$ from this distribution we use a Langevin-Hastings algorithm, the *reparametrisation* $S = F^{\mathrm{T}} m_b + S_\sigma (R + FV_b F^{\mathrm{T}}) \Omega^{1/2} \Gamma$, where $\Omega = S_\sigma^2 (R + FV_b F^{\mathrm{T}})$, and a priori $\Gamma \sim t_{n+n_\sigma}(0, I_n)$. The gradient $\nabla(\gamma)$ which determines the mean of the proposal distribution has the following form when h is the canonical link function,

$$\nabla(\gamma) = \frac{\partial}{\partial \gamma} \log f(\gamma \mid y) = -\gamma(n + n_\sigma)/(n_\sigma + \|\gamma\|^2) + (\Omega^{1/2})^{\mathrm{T}} \{y - h^{-1}(s)\}. \tag{2.14}$$

By using a conjugate prior for (β, σ^2) we find that $[\beta, \sigma^2 \mid s(j)]$, $j = 1, \ldots, m$ are Gaussian-Scaled-Inverse-χ^2 distributions with means and variances given by (2.6). From this we can calculate the mean and the variance of the posterior $[\beta, \sigma^2 \mid y]$, and also simulate from it.

Procedures similar to the ones given in Section 2.9.1 can be used for prediction. The only difference is that from (2.8), we see that $[S^* | s(j)]$, $j = 1, \ldots, m$, are now multivariate-t distributed rather than multivariate Gaussian.

Concerning the use of flat or non-informative priors for β and σ^2 in a GLSM, a word of caution is needed. The prior $1/\sigma^2$ for σ^2, recommended as a non-informative prior for the Bayesian linear Gaussian model in Section 2.7, results in an improper posterior distribution for a GLMM (see Natarajan and Kass 2000), and should therefore be avoided. Since a linear Gaussian model with a fixed positive measurement error $\tau_0^2 > 0$ can be considered as a special case of a GLSM, this is also true for such a model. There seems to be no consensus concerning reference priors for GLMM's.

We now allow for uncertainty also in ϕ, and adopting as our prior

$$\pi(\beta, \sigma^2, \phi) = \pi_{\{N\chi^2_{ScI}\}}(\beta, \sigma^2)\pi(\phi),$$

where $\pi(\phi)$ is any proper prior.

When using an MCMC-algorithm updating ϕ, we need to calculate $(R(\phi) + FV_b F^{\mathrm{T}})^{1/2}$ for each new ϕ value, which is the most time-consuming part of the algorithm. To avoid this significant increase in computation time, we adopt a discrete prior for ϕ on a set of values covering the range of interest, and precompute and store $(R(\phi) + FV_b F^{\mathrm{T}})^{1/2}$ for each value of ϕ.

To simulate from $[S, \phi | y]$, after integrating out β and σ^2, we use a hybrid Metropolis-Hastings algorithm where S and ϕ are updated sequentially. The update of S is of the same type as used earlier, with ϕ equal to the present value in the MCMC iteration. To update ϕ we use a random walk

Metropolis update where the proposal distribution is a Gaussian distribution rounded to the nearest ϕ value in the discrete set for the prior. The output of this algorithm is a sample $(s(1), \phi(1)), \ldots, (s(m), \phi(m))$ from the distribution $[S, \phi \mid y]$.

The predictive distribution for S^* is given by

$$p(s^*|y) = \int\int p(s^* \mid s, \phi) p(s, \phi \mid y) ds d\phi$$

To simulate from this predictive distribution, we simulate $s^*(j)$ from $[S^* \mid s(j), \phi(j)]$, which is multivariate-t, $j = 1, \ldots, m$.

We may also want to introduce a nugget term into the specification of the model, replacing $S(x_i)$ by $S(x_i)+U_i$ where the U_i are mutually independent Gaussian variates with mean zero and variance τ^2. Here, in contrast to the Gaussian case, we can make a formal distinction between the U_i as a representation of micro-scale variation and the error distribution induced by the sampling mechanism, for example Poisson for count data. In some contexts, the U_i may have a more specific interpretation. For example, if a binary response were obtained from each of a number of sampling units at each of a number of locations, a binomial error distribution would be a natural choice, and the U_i and $S(x_i)$ would then represent, respectively, non-spatial and spatial sources of extra-binomial variation. The inferential procedure is essentially unchanged, except that we now use a discrete joint prior for (ϕ, τ^2).

2.9.3 A spatial model for count data

A GLSM for modelling spatial count data is the Poisson-log-linear spatial model, in which $[Y_i \mid S(x_i)]$ follows a Poisson distribution with mean $t_i \exp(S(x_i))$, $i = 1, \ldots, n$. The term t_i may, for example, represent a time-interval over which the corresponding count Y_i is accumulated, as in Diggle et al. (1998), or an area within which the number of events Y_i is counted, as in Christensen & Waagepetersen (2002).

We assume initially that parameters are known, and that we are interested in predicting the intensity $\lambda(x_0) = \exp(S(x_0))$ at a location x_0. Given a sample $s(1), \ldots, s(m)$ from $[S|y]$, obtained using the MCMC-algorithm in Section 2.9.1, $[S(x_0)|s(j)], j = 1, ..., m$ follow multivariate Gaussian distributions. Since the moments of the exponential of a multivariate Gaussian distribution are obtainable in closed form, the following procedure can be used for predicting $\lambda(x_0) = \exp(S(x_0))$.

- Calculate $\mathrm{E}[S(x_0)|s(j)]$ and $\mathrm{Var}[S(x_0)|s(j)]$, $j = 1, \ldots, m$, using kriging.
- Calculate, for each of $j = 1, \ldots, m$,

$$\mathrm{E}[\lambda(x_0)|s(j)] = \exp(\mathrm{E}[S(x_0)|s(j)] + 0.5\mathrm{Var}[S(x_0)|s(j)])$$

- Approximate

$$\mathrm{E}[\lambda(x_0)|y] \approx \frac{1}{m}\sum_{j=0}^{m} \mathrm{E}[\lambda(x_0)|s(j)]$$

Note that $\mathrm{E}[\exp(\alpha S)|y]$ is finite for any $\alpha \in \mathbb{R}^n$, $\mathrm{E}[S(x_0)|S]$ is a linear function of S, and $\mathrm{Var}[S(x_0)|S]$ does not depend on S. Therefore $\mathrm{E}[\lambda(x_0)|y] < \infty$, and the quantity we want to approximate using MCMC exists. As we shall see below, if we use only simulation-based methods we may unwittingly produce estimates of quantities that do not exist.

An algorithm similar to one above could, in principle, be used when we want to incorporate prior information into the predictions by using the conjugate Gaussian-Scaled-Inverse-χ^2 prior for (β, σ^2), the difference being that $[S(x_0) \mid s(j)]$, $j = 1, \ldots, m$ are now multivariate-t distributions. However, because the mean of the exponential of a multivariate-t distribution is not finite, the procedure fails. In fact, the minimum mean square error predictor does not exist in this case. Had we used a different MCMC-algorithm, sampling β and σ^2 instead of integrating them out, or had we decided to generate a sample $\exp(s_0(1)), \ldots, \exp(s_0(m))$ instead of using the formula for $\mathrm{E}[\exp(S(x_0))|s(j)]$, $j = 1, \ldots, m$, this problem might have been missed. This method would, of course, have generated a valid sample from the required predictive distribution. If we do want to quote a point prediction in a situation of this kind, we might for example use the predictive median rather than the mean.

2.9.4 Spatial model for binomial data

A GLSM for binomial data is as follows. The data are arranged as triples, (x_i, y_i, n_i), where y_i is a count of the number of successes out of n_i Bernoulli trials associated with the location x_i. Conditional on an unobserved Gaussian process $S(\cdot)$, we model the y_i as realisations of mutually independent binomial random variables with numbers of trials n_i and success probabilities $p_i = p(x_i)$, where

$$\log\{p(x)/(1 - p(x))\} = S(x). \tag{2.15}$$

As before, the process $S(\cdot)$ has spatially varying mean $\mu(x) = \sum f_j(x)\beta_j$, variance σ^2 and correlation parameter ϕ.

To illustrate the prediction problem in this context, suppose that the target for prediction is $T = p(x_0)$. Because no closed form expressions can be found for the mean and variance of $[T \mid S]$ we need to simulate from this distribution. Assuming a Gaussian-Scaled-Inverse-χ^2 prior for (β, σ^2), and a proper prior for ϕ, we proceed as follows:

- simulate $((s(1), \phi(1)), \ldots, (s(m), \phi(m))$ from $[S, \phi|y]$, using MCMC;

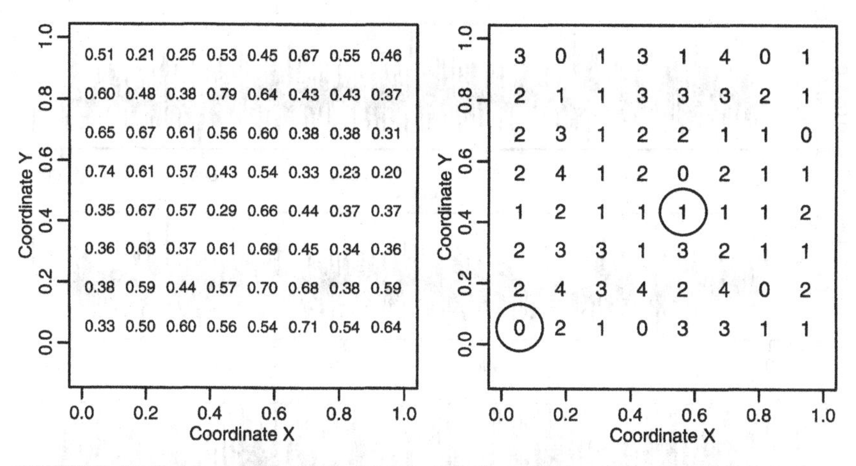

FIGURE 2.17. Map of simulated binomial data. Left: the values of underlying process $S(\cdot)$. Right: binomial data with size 4 and probability parameter equal to the inverse logit-function; circles indicates locations for which MCMC traces will be shown.

- calculate $\mathrm{E}[S(x_0)|s(j), \phi(j)]$ and $\mathrm{Var}[S(x_0)|s(j), \phi(j)]$ for each of $j = 1, \dots, m$;
- simulate values $s_0(j), j = 1, ..., m$ from multivariate-t distributions with common degrees of freedom $n + n_\sigma$, means $\mathrm{E}[S(x_0)|s(j)]$ and variances $\mathrm{Var}[S(x_0)|s(j)]$;
- approximate

$$\mathrm{E}[T|y] \approx \frac{1}{m} \sum_{j=0}^{m} \exp(s_0(j))/(1 + \exp(s_0(j))).$$

Heagerty & Lele (1998) and De Oliveira (2000) use an apparently different model for spatial binary data which they call the clipped Gaussian field. In this model, the measurement process is $\{Y(x) = 1_{\{S(x)>0\}} : x \in A\}$, where $S(\cdot)$ is a Gaussian process. Assuming that the process $S(\cdot)$ has a positive nugget τ^2, we can write this model as $Y(x) = 1_{\{\tilde{S}(x)+U(x)>0\}}$, where $\tilde{S}(\cdot)$ is another Gaussian process and $U(\cdot)$ is a Gaussian white noise process with mean 0 and variance 1. The conditional distribution $[Y(x) \mid \tilde{S}(x)]$ is binomial of size 1 and probability $P(Y(x) = 1 \mid \tilde{S}(x)) = \Phi(\tilde{S}(x))$. The model is therefore identical to the one described above, except that the logit link in (2.15) is replaced by the probit link.

2.9.5 Example

To illustrate the inferential procedure in a GLSM we consider the simulated data-set shown in Figure 2.17 which consists of binomial data at 64

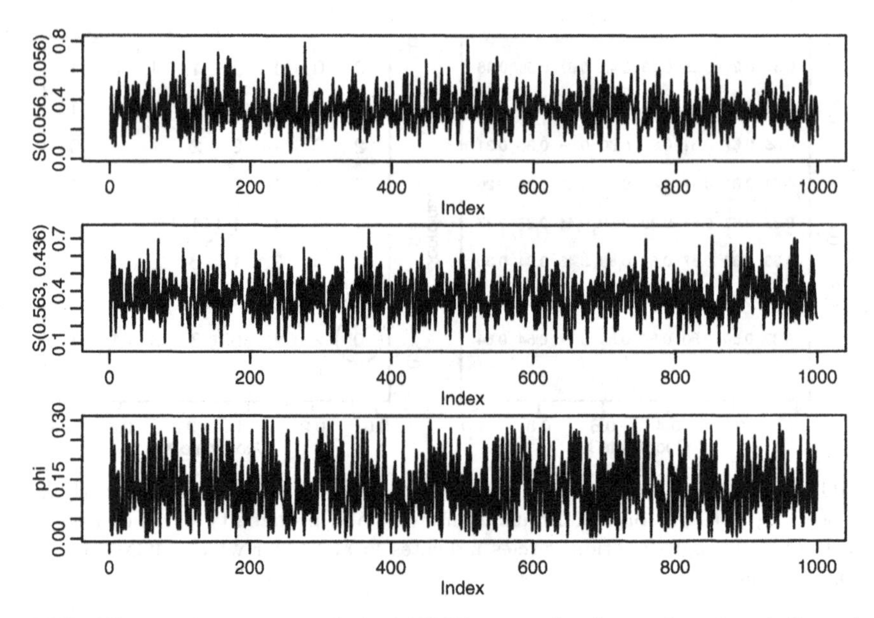

FIGURE 2.18. Time series with the MCMC output for the two locations indicated by circles in Figure 2.17 and for the parameter ϕ.

locations. The left-hand panel shows the values of the underlying Gaussian random variables $S(x_1), \ldots, S(x_n)$ simulated from a model with exponential correlation function and parameter values $(\beta, \sigma^2, \phi) = (0, 0.5, 0.2)$. The right-hand plot shows the corresponding binomial variables which are simulated from independent binomial distributions of size 4 and probability parameter $p(x_i) = \exp(S(x_i))/(1 + \exp(S(x_i)))$ $i = 1, \ldots, n$, the inverse logit transform of $S(\cdot)$. Note that inference is based on the observed binomial values in the right hand plot, with $S(x_1), \ldots, S(x_n)$ in the left hand plot considered as unobserved.

We perform a Bayesian analysis with exponential correlation function and the following priors: a Gaussian $N(0, \sigma^2)$ prior for $[\beta \mid \sigma^2]$, a $\chi^2_{ScI}(5, 0.5)$ prior for σ^2, and a discrete exponential prior for ϕ, $\pi(\phi) = \exp(-\phi/0.2)$, with 60 discretisation points in the interval $[0.005, 0.3]$.

For inference we run the MCMC-algorithm described in Section 2.9.2, discarding the first 10,000 iterations then retaining every 100th of 100,000 iterations to obtain a sample of size 1000. Figure 2.18 shows the output for the two S coordinates circled in Figure 2.17, and for the parameter ϕ. The estimated autocorrelations are in each case less than 0.1 for all positive lags, and the thinned sample has very low autocorrelation.

For prediction of the probabilities over the area, we consider 1600 locations in a regular square grid and use the procedure described in Section 2.9.4. The left-hand panel of Figure 2.19 shows the predictions at the 1600 locations, whilst the right-hand panel shows the associated prediction

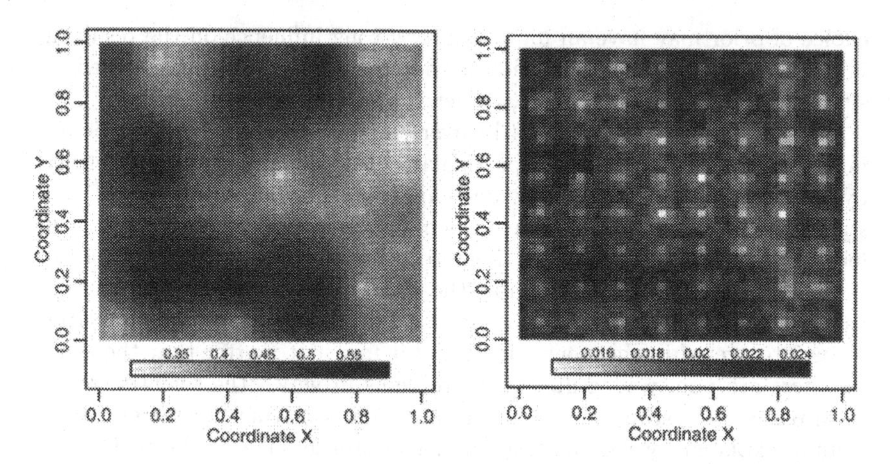

FIGURE 2.19. Left: predicted values at the grid points. Right: prediction variances.

variances.

Comparing the left-hand panels of Figure 2.17 and Figure 2.19 we see that the predicted surface has less extreme values than does the true surface $S(\cdot)$, as is to be expected when predicting from noisy data. The prediction variances on the right-hand plot in Figure 2.19 show a weak dependence on the means, with a preponderance of large values of the prediction variance in areas where the predicted means are close to 0.5. The effect of the sampling design is also clear, with small prediction variances at locations close to grid-points.

2.10 Discussion

In this short introduction to the subject of model-based geostatistics, our aim has been to set out the basic methodology for dealing with geostatistical problems from the perspective of mainstream parametric statistical modelling. Under Gaussian assumptions, the resulting prediction methodology has a very close relationship to classical geostatistical kriging methods, but the treatment of unknown parameters departs markedly from the classical geostatistical approach. The classical approach uses curve-fitting methods to match empirical and theoretical variograms, whereas we have emphasised the use of the likelihood function for parameter estimation, whether from a Bayesian or non-Bayesian point of view. The Bayesian approach has the attractive property that uncertainty in the values of model parameters is recognised in the construction of prediction intervals, leading to a more honest assessment of prediction error. This should not blind us to the uncomfortable fact that even the simplest geostatistical models may be poorly identifiable from the available data, and in these situations the choice of prior may have an unpleasantly strong influence on the resulting inferences. However, our suspicion is that more ad hoc methods based on simple plug-in methods conceal, rather than solve, this difficulty.

One general point that cannot be over-emphasised is that the model and inferential method used on a particular set of data should be informed by the scientific objective of the data analysis. The full machinery of geostatistical modelling is indicated when prediction at unobserved spatial locations is a central objective. For problems of this kind, some or all of the model parameters, for example those defining the assumed spatial covariance structure of the data, are means to an end rather than quantities to be interpreted in their own right, and it may not matter that these parameters are poorly identified. For problems in which the scientific focus is on parameter estimation, for example in investigating the regression effects of explanatory variables, a simpler approach such as the method of generalised estimating equations may be all that is required (Gotway & Stroup 1997). Note, however that this changes the interpretation of the regression parameter as affecting marginal, rather than conditional expectations.

We acknowledge that our use of the Bayesian inferential paradigm is pragmatic. We find it difficult to come up with convincing arguments for the choice of priors in a geostatistical model, but we do want a prediction methodology which acknowledges all of the major sources of uncertainty in our predictions.

We have omitted altogether a number of topics due to restrictions on space. Within the linear Gaussian setting, extensions of the methodology to multivariate and/or space-time data are straightforward in principle, although the enriched data-structure leads to a proliferation of modelling choices. Some examples are included in the suggestions given in Section 2.12 for further reading. Also, for large space-time data-sets, apparently obvious approaches may be computationally infeasible.

Outside the generalised linear model setting, the number of potential models is practically limitless and the ability to fit by Monte Carlo methods almost arbitrarily complex models is a two-edged sword. On one hand, it is right in principle that models should be informed by scientific knowledge, rather than chosen from an artificially restricted class of analytically or numerically tractable models. Against this, it is all too easy to devise a model whose complexity far outstrips the capacity of the data to provide reliable validation of its underlying assumptions.

A fundamental problem which is ignored in most geostatistical work is the possibility of stochastic interaction between the signal or measurement process and the sampling design. For example, in mineral exploration samples will be taken from locations which are thought likely to yield commercially viable grades of ore. The formal framework for handling problems of this kind is a *marked point process*, a joint probability model for a stochastic *point process* X, and an associated set of random variables, or *marks*,

Y (see Section 4.3.2). As always, different factorisations of the joint distribution are available, and whilst these factorisations are mathematically equivalent, in practice they lead to different modelling assumptions. The simplest structural assumption is that X and Y are independent processes, hence $[X, Y] = [X][Y]$. This is sometimes called the *random field model*, and is often assumed implicitly in geostatistical work. In the dependent case, one possible factorisation is $[X, Y] = [X|Y][Y]$. This is a natural way to describe what is sometimes called *preferential sampling*, in which sampling locations are determined by partial knowledge of the underlying mark process; an example would be the deliberate siting of air pollution monitors in badly polluted areas. The opposite factorisation, $[X, Y] = [X][Y|X]$, may be more appropriate when the mark process is only defined at the sampling locations; for example, the heights of trees in a naturally regenerated forest. The full inferential implications of ignoring violations of the random field model have not been widely studied.

2.11 Software

All the analyses reported in this chapter have been carried out using the packages `geoR` and `geoRglm`, both of which are add-on's to the freely available and open-source statistical system `R` (Ihaka & Gentleman 1996). The official web site of the `R`-project is at *www.r-project.org*. Both packages are available in the contributed section of `CRAN` (Comprehensive R Archive Network).

The package `geoR` (Ribeiro Jr & Diggle 2001) implements basic geostatistical tools and the methods for Gaussian linear models described here. Its official web site is *www.maths.lancs.ac.uk/~ribeiro/geoR*, where instructions for downloading and installation can be found, together with a tutorial on the package usage.

The package `geoRglm` is an extension of `geoR` which implements the generalised linear spatial model described in Section 2.9. It is available at *www.lancaster.ac.uk/~christen/geoRglm*, together with an introduction to the package.

Other computational resources for analysis of spatial data using `R` are reviewed in issues 2 and 3 of *R-NEWS*, available at *cran.r-project.org/doc/Rnews*. An extensive collection of geostatistics materials can be found in the AI-GEOSTATS web site at *www.ai-geostats.org*.

2.12 Further reading

Chilès & Delfiner (1999) is a standard reference for classical geostatistical methods. Cressie (1993) describes geostatistics as one of three main branches of spatial statistics. Stein (1999) gives a rigorous account of the mathematical theory underlying linear kriging. Ribeiro Jr & Diggle (1999) give a more detailed presentation of Bayesian inference for the linear Gaussian model. Other references to Bayesian inference for geostatistical models include Kitanidis (1986), Le & Zidek (1992), Handcock & Stein (1993), De Oliveira et al. (1997), Diggle & Ribeiro Jr (2002), De Oliveira & Ecker (2002) and Berger et al. (2001). Omre & Halvorsen (1989) describe the link between Bayesian prediction and simple or ordinary kriging.

Christensen, Møller & Waagepetersen (2001) give further details and properties of the Langevin-Hastings algorithm used in Section 2.9.1. Bayesian inference for a GLSM is described in Diggle et al. (1998) and in Christensen & Waagepetersen (2002) where algorithms are used that update both the random effect S and all the parameters (including β and σ^2). These algorithms are more general than the one presented in this chapter, since they do not require conjugate priors (or limiting cases of conjugate priors). However, from our experience, in practice a consequence of the extra generality is that the algorithms need to be more carefully tuned to specific applications in order to achieve good mixing properties. Zhang (2002) analyses spatial binomial data, and develops a Monte Carlo EM gradient algorithm for maximum likelihood estimation in a GLSM.

Multivariate spatial prediction is presented in Chapter 5 in Chilés & Delfiner (1999); see also Brown, Le & Zidek (1994), Le, Sun & Zidek (1997) and Zidek, Sun & Le (2000).

Examples of space-time modelling include Handcock & Wallis (1994), Wikle & Cressie (1999), Brown, Kårensen, Roberts & Tonellato (2000), Brix & Diggle (2001) and Brown, Diggle, Lord & Young (2001).

Wälder & Stoyan (1996), Wälder & Stoyan (1998) and Schlather (2001) discuss the connection between the classical variogram and the more general second-order properties of marked point processes. Marked point processes are also discussed in Chapters 3 and 4.

Acknowledgments: We thank the UK Engineering and Physical Sciences Research Council and CAPES (Brasil) for financial support.

2.13 References

Adler, R. J. (1981). *The Geometry of Random Fields*, Wiley, New York.

Berger, J. O., De Oliveira, V. & Sansó, B. (2001). Objective Bayesian analysis of spatially correlated data, *Journal of the American Statistical Association* **96**: 1361–1374.

Besag, J., York, J. & Mollié, A. (1991). Bayesian image restoration, with two applications in spatial statistics (with discussion), *Annals of the Institute of Statistical Mathematics* **43**: 1–59.

Breslow, N. E. & Clayton, D. G. (1993). Approximate inference in generalized linear mixed models, *Journal of the American Statistical Association* **88**: 9–25.

Brix, A. & Diggle, P. J. (2001). Spatio-temporal prediction for log-Gaussian Cox processes, *Journal of the Royal Statistical Society, Series* B **63**: 823–841.

Brown, P. E., Diggle, P. J., Lord, M. E. & Young, P. C. (2001). Space-time calibration of radar rainfall data, *Applied Statistics* **50**: 221–241.

Brown, P. E., Kårensen, K. F., Roberts, G. O. & Tonellato, S. (2000). Blur-generated non-separable space-time models, *Journal of the Royal Statistical Society, Series* B **62**: 847–860.

Brown, P. J., Le, N. D. & Zidek, J. V. (1994). Multivariate spatial interpolation and exposure to air polutants, *Canadian Journal of Statistics* **22**: 489–509.

Chilés, J. P. & Delfiner, P. (1999). *Geostatistics; Modeling Spatial Uncertainty*, Wiley, New York.

Christensen, O. F., Diggle, P. J. & Ribeiro Jr, P. J. (2001). Analysing positive-valued spatial data: the transformed Gaussian model, *in* P. Monestiez, D. Allard & R. Froidevaux (eds), *GeoENV III - Geostatistics for Environmental Applications*, Vol. 11 of *Quantitative Geology and Geostatistics*, Kluwer, Dordrecht, pp. 287–298.

Christensen, O. F., Møller, J. & Waagepetersen, R. (2001). Geometric ergodicity of Metropolis Hastings algorithms for conditional simulation in generalised linear mixed models, *Methodology and Computing in Applied Probability* **3**: 309–327.

Christensen, O. F. & Waagepetersen, R. P. (2002). Bayesian prediction of spatial count data using generalized linear mixed models, *Biometrics* **58**: 280–286.

Cramér, H. & Leadbetter, M. R. (1967). *Stationary and Related Processes*, Wiley, New York.

Cressie, N. (1993). *Statistics for Spatial Data – revised edition*, Wiley, New York.

De Oliveira, V. (2000). Bayesian prediction of clipped Gaussian random fields, *Computational Statistics and Data Analysis* **34**: 299–314.

De Oliveira, V. & Ecker, M. D. (2002). Bayesian hot spot detection in the presence of a spatial trend: application to total nitrogen concentration in the Chesapeake Bay, *Environmetrics* **13**: 85–101.

De Oliveira, V., Kedem, B. & Short, D. A. (1997). Bayesian prediction of transformed Gaussian random fields, *Journal of the American Statistical Association* **92**: 1422–1433.

Diggle, P. J., Harper, L. & Simon, S. (1997). Geostatistical analysis of residual contamination from nuclear weapons testing, *in* V. Barnett & F. Turkman (eds), *Statistics for the Environment 3: Pollution Assesment and Control*, Wiley, Chichester, pp. 89–107.

Diggle, P. J. & Ribeiro Jr, P. J. (2002). Bayesian inference in Gaussian model based geostatistics, *Geographical and Environmental Modelling* **6**. To appear.

Diggle, P. J., Tawn, J. A. & Moyeed, R. A. (1998). Model based geostatistics (with discussion), *Applied Statistics* **47**: 299–350.

Geyer, C. J. (1992). Practical Markov chain Monte Carlo (with discussion), *Statistical Science* **7**: 473–511.

Goovaerts, P. (1997). *Geostatistics for Natural Resources Evaluation*, Oxford University Press, New York.

Gotway, C. A. & Stroup, W. W. (1997). A generalized linear model approach to spatial data analysis and prediction, *Journal of Agricultural, Biological and Environmental Statistics* **2**: 157–178.

Handcock, M. S. & Stein, M. L. (1993). A Bayesian analysis of kriging, *Technometrics* **35**: 403–410.

Handcock, M. S. & Wallis, J. R. (1994). An approach to statistical spatial-temporal modeling of meteorological fields, *Journal of the American Statistical Association* **89**: 368–390.

Heagerty, P. J. & Lele, S. R. (1998). A composite likelihood approach to binary spatial data, *Journal of the American Statistical Association* **93**: 1099–1111.

Ihaka, R. & Gentleman, R. (1996). R: A language for data analysis and graphics, *Journal of Computatioanl and Graphical Statistics* **5**: 299–314.

Journel, A. G. & Huijbregts, C. J. (1978). *Mining Geostatistics*, Academic Press, London.

Kent, J. T. (1989). Continuity properties of random fields, *Annals of Probability* **17**: 1432–1440.

Kitanidis, P. K. (1983). Statistical estimation of polynomial generalized covariance functions and hydrological applications., *Water Resources Research* **22**: 499–507.

Kitanidis, P. K. (1986). Parameter uncertainty in estimation of spatial functions: Bayesian analysis, *Water Resources Research* **22**: 499–507.

Krige, D. G. (1951). A statistical approach to some basic mine valuation problems on the Witwatersrand, *Journal of the Chemical, Metallurgical and Mining Society of South Africa* **52**: 119–139.

Le, N. D., Sun, W. & Zidek, J. V. (1997). Bayesian multivariate spatial interpolation with data missing by design, *Journal of the Royal Statistical Society, Series* B **59**: 501–510.

Le, N. D. & Zidek, J. V. (1992). Interpolation with uncertain covariances: a Bayesian alternative to kriging, *Journal of Multivariate Analysis* **43**: 351–374.

Mardia, K. V. & Watkins, A. J. (1989). On multimodality of the likelihood in the spatial linear model, *Biometrika* **76**: 289–296.

Matérn, B. (1960). Spatial Variation. Meddelanden fran Statens Skogforskningsinstitut, Band 49, No. 5.

McCullagh, P. & Nelder, J. A. (1989). *Generalized Linear Models*, second edn, Chapman and Hall, London.

Natarajan, R. & Kass, R. E. (2000). Bayesian methods for generalized linear mixed models, *Journal of the American Statistical Association* **95**: 227–237.

O'Hagan, A. (1994). *Bayesian Inference*, Vol. 2b of *Kendall's Advanced Theory of Statistics*, Edward Arnold.

Omre, H. & Halvorsen, K. B. (1989). The Bayesian bridge between simple and universal kriging, *Mathematical Geology* **21**: 767–786.

Patterson, H. D. & Thompson, R. (1971). Recovery of inter-block information when block sizes are unequal, *Biometrika* **58**: 545–554.

Ribeiro Jr, P. J. & Diggle, P. J. (1999). Bayesian inference in Gaussian model-based geostatistics, *Tech. Report ST-99-09*, Lancaster University.

Ribeiro Jr, P. J. & Diggle, P. J. (2001). geoR: a package for geostatistical analysis, *R News* **1/2**: 15–18. Available from: http://www.r-project.org/doc/Rnews.

Ripley, B. D. (1981). *Spatial Statistics*, Chapman and Hall, New York.

Sampson, P. D. & Guttorp, P. (1992). Nonparametric estimation of nonstationary spatial covariance structure, *Journal of the American Statistical Association* **87**: 108–119.

Schlather, M. (2001). On the second-order characteristics of marked point processes, *Bernoulli* **7**: 99–117.

Stein, M. L. (1999). *Interpolation of Spatial Data: Some Theory for Kriging*, Springer Verlag, New York.

Wälder, O. & Stoyan, D. (1996). On variograms in point process statistics, *Biometrical Journal* **38**: 895–905.

Wälder, O. & Stoyan, D. (1998). On variograms in point process statistics: Erratum, *Biometrical Journal* **40**: 109.

Warnes, J. J. & Ripley, B. D. (1987). Problems with likelihood estimation of covariance functions of spatial Gaussian processes, *Biometrika* **74**: 640–642.

Whittle, P. (1954). On stationary processes in the plane, *Biometrika* **41**: 434–449.

Whittle, P. (1962). Topographic correlation, power-law covariance functions, and diffusion, *Biometrika* **49**: 305–314.

Whittle, P. (1963). Stochastic processes in several dimensions, *Bulletin of the International Statistical Institute* **40**: 974–994.

Wikle, C. K. & Cressie, N. (1999). A dimension-reduced approach to space-time Kalman filtering, *Biometrika* **86**: 815–829.

Zhang, H. (2002). On estimation and prediction for spatial generalised linear mixed models, *Biometrics* **58**: 129–136.

Zidek, J. V., Sun, W. & Le, N. D. (2000). Designing and integrating composite networks for monitoring multivariate Gaussian pollution field, *Applied Statistics* **49**: 63–79.

Zimmerman, D. L. (1989). Computationally efficient restricted maximum likelihood estimation of generalized covariance functions, *Mathematical Geology* **21**: 655–672.

3

A Tutorial on Image Analysis

Merrilee A. Hurn
Oddvar K. Husby
Håvard Rue

3.1 Introduction

3.1.1 Aims of image analysis

Data arise in the form of images in many different areas and using many different technologies. Within medical diagnostics, X-rays are probably the most well-known form of direct imaging, gathering structural information about the body by recording the transmission of X-rays. More recent advances have been the various emission-based technologies, PET and SPECT, which aim to map metabolic activity in the body, and MRI (Figure 3.1c) which again provides structural information. On two quite different scales, satellites (Figure 3.1a and b) and microscopes (Figure 3.1d) monitor and record useful scientific information; for example, aerial imaging in different wavebands and at different stages in the growing season can be used to detect crop subsidy fraud, while some types of microscopy can be used to generate temporal sequences of three dimensional images, leading to a greater understanding of biological processes. There is an expectation that technological advances should soon provide solutions to problems such as automatic face or hand recognition, or unsupervised robotic vision.

A wide range of image processing tasks arise, not merely because of the range of different applications, but also because of the diversity of goals. In the medical context, some forms of imaging, such as PET, involve the use of radioactive material, and so the exposure should ideally be as small as possible. However this is a compromise with maintaining good image quality, and so the processing question may be one of making the visual appearance of the image as clear as possible. In other applications, it may not be the picture quality which is so much at issue as the extraction of quantitative information. We may be faced with questions ranging from "Can you make this picture less blurred" to "Can you describe the packing structure of any cells in this medium". Generally tasks of the former type, trying to improve the image quality, require local actions (deconvolution, or noise removal for example), and are known as *low level tasks.* Tasks which address more global properties of the scene, such as locating or identifying

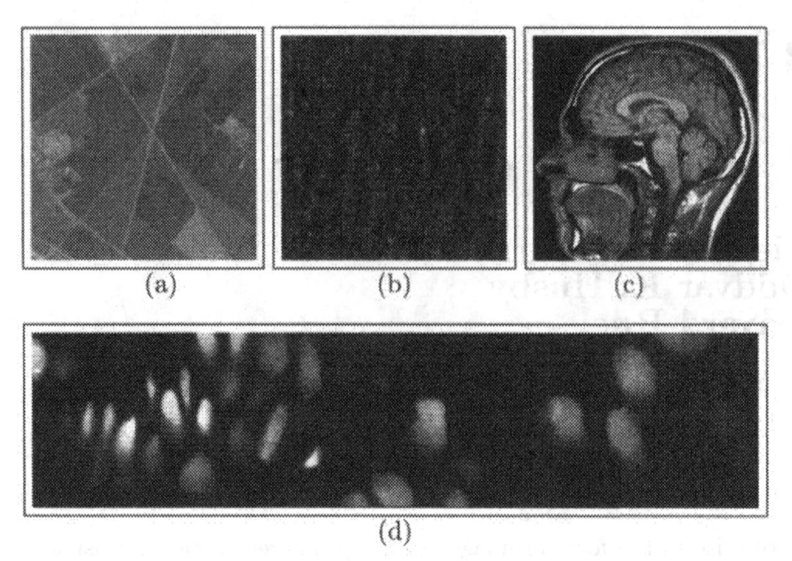

FIGURE 3.1. Some examples of image data; (a) A satellite image over an agricultural area, (b) a satellite image of ocean waves, (c) an MRI image of the brain, and (d) a confocal microscopy image of cells.

objects, are referred to as *high-level tasks.*

One unifying theme for the different types of imaging problems is that they may generally be regarded as being of the forms

$$\text{Signal} = \text{Image} \otimes \text{Noise}, \tag{3.1}$$

or

$$\text{Signal} = f(\text{Image}) \otimes \text{Noise}, \tag{3.2}$$

where $\otimes$ represents a suitable combination operator, and the function f indicates that the signal is not of the same format as the original image (for example, in emission tomography, the signals are emission counts along lines through the body). Notice that we are being rather vague about what the image actually is; this is a point to which we will return when we discuss modelling for different tasks.

3.1.2 Bayesian approach

What role can Bayesian statistics play in image analysis and processing? This is a particularly pertinent question given that there are many quick, often deterministic, algorithms developed in the computer science and electrical engineering fields. As we see it, there are three main contributions: The first is to model the noise structure adequately; the second is to regularise underdetermined systems through the use of a prior distribution; and the final, possibly most important role, is to be able to provide confidence statements about the output results. Increasingly, with the advance in technology, the goals of the imaging are more complex than before and the questions being posed are more quantitative than qualitative. This is particularly an issue when studies also have a temporal aspect where it is important to be able to separate real change from noise.

In these notes, we will denote the underlying image by $\boldsymbol{x}$, although for the moment we postpone saying exactly what $\boldsymbol{x}$ is, and the corresponding signal by $\boldsymbol{y}$. Often, but not always, $\boldsymbol{y}$ will be a lattice of discrete values; in which case, the individual components of $\boldsymbol{y}$ are known as pixels (short for picture element). Bayes theorem allows us to write

$$\pi(\boldsymbol{x} \mid \boldsymbol{y}) = \frac{\pi(\boldsymbol{y} \mid \boldsymbol{x})\pi(\boldsymbol{x})}{\pi(\boldsymbol{y})} \propto \pi(\boldsymbol{y} \mid \boldsymbol{x})\pi(\boldsymbol{x}) \tag{3.3}$$

where the likelihood, $\pi(\boldsymbol{y}|\boldsymbol{x})$, describes the data formation process for a particular underlying image, while the prior, $\pi(\boldsymbol{x})$, encodes any prior beliefs about the properties of such underlying images. The marginal for the data $\pi(\boldsymbol{y})$ is uninformative regarding $\boldsymbol{x}$. We will of course be interested in drawing inferences about $\boldsymbol{x}$ based on the posterior $\pi(\boldsymbol{x}|\boldsymbol{y})$. To do this we will need to specify various components of the model. First we must decide what a suitable representational form is for $\boldsymbol{x}$, and to a large extent this will depend on what the objectives of the imaging are. We must then decide upon appropriate prior and likelihood models. In the following two sections we will discuss some modelling possibilities, and also describe associated issues of the treatment of nuisance parameters and the machinery for inference.

3.1.3 Further reading

As this chapter is a tutorial and not a research paper, we have been relatively relaxed about referring to all the appropriate sources for some of the more basic material. Further, we do not claim to give a complete, or even almost complete, overview of this huge field. There are many topics which could have been discussed but which are not mentioned at all. For the reader who wishes to learn more about this field, there are several good sources. A good start might be the RSS discussion papers (Besag 1974, Besag 1986, Grenander & Miller 1994, Glasbey & Mardia 2001) (one in each decade), Geman & Geman (1984) and Geman (1990) are essential reading, as is Grenander (1993), although this might be a bit tough. The books by Winkler (1995) and Dryden & Mardia (1999) are also worth reading. Nice applications of high-level models can also be found in Blake & Isard (1998).

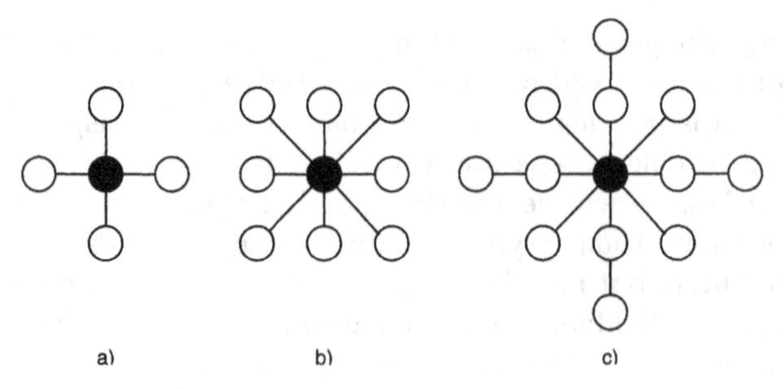

FIGURE 3.2. Common neighbourhood structures in imaging, four, eight or twelve nearest neighbours.

3.2 Markov random field models

We will begin with models where the representation for $\boldsymbol{x}$ is as that for $\boldsymbol{y}$, that is as a list of pixel values. We will need some definitions: Let $\mathcal{I}$ be the set of sites or pixels in the image which is assumed to be finite. Each $i \in \mathcal{I}$ is a coordinate in the lattice. In order to define a pixel-based model for images, we turn to the class of *Markov random fields*. We begin by defining a symmetric neighbourhood relation $\sim$ on $\mathcal{I}$, if i is a *neighbour* of j (written as $i \sim j$) then j is a neighbour of i. By convention i is not a neighbour of itself. A random field is then a collection of random variables $\{x_i : i \in \mathcal{I}\}$ where each x_i takes values in a finite set χ. Denote the neighbours of i by ∂i. For a subset of sites $A \subseteq \mathcal{I}$ we also use the notation $x_A = \{x_i : i \in A\}$ and $\boldsymbol{x}_{-A} = \{x_i : i \in \mathcal{I} \setminus A\}$. An $\boldsymbol{x}$ *configuration* is an element of $\chi^{|\mathcal{I}|}$. A random field is called a *Markov random field* (MRF) if the conditional distribution of any pixel given the rest (also called its *local characteristic*) only depends on the values of that pixel's neighbours,

$$\pi(x_i \mid \boldsymbol{x}_{-i}) = \pi(x_i \mid \boldsymbol{x}_{\partial i}). \tag{3.4}$$

Commonly used neighbourhood structures in imaging are four, eight or twelve nearest neighbours, see Figure 3.2.

MRFs are important in imaging for two main reasons:

1. Modelling the joint distribution of an image $\boldsymbol{x}$ on the lattice $\mathcal{I}$ is a daunting task because it is not immediately clear even how to begin. Approaching the issue through the *full conditional distributions* breaks the problem down into more manageable tasks, in the sense that we may be able to say more clearly how we think x_i behaves if we know the configuration of its neighbours.

2. There is an important connection between Markov chain Monte Carlo methods (MCMC) and MRFs, in that single-site updating schemes in MCMC only require evaluations of the local full conditionals (3.4). If we assume that the number of neighbours is very considerably

less than $n = |\mathcal{I}|$, as is the case for almost all applications, then a full sweep of the image using a Gibbs sampler, say, requires $\mathcal{O}(n)$ operations for a MRF, as opposed to $\mathcal{O}(n^2)$ operations for a random field lacking this local property.

Although we might therefore approach the modelling problem by specifying a neighbourhood and the full conditionals for each site, one very important issue arises: Given a set of local characteristics, under what conditions are we guaranteed that a legitimate joint density exists? Is this joint density unique and what is it? These are delicate questions. Suppose we have a set of full conditionals, and we wish to construct a corresponding joint density. Since the joint density sums to 1, it is enough to study $\pi(\boldsymbol{x})/\pi(\boldsymbol{x}^*)$, for some reference configuration $\boldsymbol{x}^*$. By considering cancellation of successive terms (and assuming $\pi(\cdot) > 0$), this can be written

$$\begin{aligned}\frac{\pi(\boldsymbol{x})}{\pi(\boldsymbol{x}^*)} &= \prod_{i=0}^{n-1} \frac{\pi(x_1^*, \ldots, x_i^*, x_{i+1}, x_{i+2}, \ldots, x_n)}{\pi(x_1^*, \ldots, x_i^*, x_{i+1}^*, x_{i+2}, \ldots, x_n)} \\ &= \prod_{i=0}^{n-1} \frac{\pi(x_{i+1} | x_1^*, \ldots, x_i^*, x_{i+2}, \ldots, x_n)}{\pi(x_{i+1}^* | x_1^*, \ldots, x_i^*, x_{i+2}, \ldots, x_n)}. \end{aligned} \tag{3.5}$$

Hence, we can obtain the joint density from a product of ratios of full conditionals, after renormalisation. Note that only the neighbours are needed in the above conditioning, but this we ignore for notational convenience. For a set of full conditionals to define a legitimate joint density, we must ensure that the joint density as defined in (3.5) is invariant to the ordering of the indices, and further, is invariant to the choice of the reference state $\boldsymbol{x}^*$. These are the consistency requirements on the full conditionals. Although it is possible, in theory, to verify these consistency requirements directly, we nearly always make use of the *Hammersley-Clifford theorem.* This theorem states that a set of full conditionals defines a legitimate joint density if and only if they are derived from a joint density of a particular form.

Theorem 1 (Hammersley-Clifford) *A distribution satisfying $\pi(\boldsymbol{x}) > 0$ for all configurations in $\chi^{|\mathcal{I}|}$ is a Markov random field if, and only if, it has a joint density of the form*

$$\pi(\boldsymbol{x}) = \frac{1}{Z} \exp\left(- \sum_{C \in \mathcal{C}} \Phi_C(\boldsymbol{x}_C) \right) \tag{3.6}$$

for some functions $\{\Phi_C\}$, where $\mathcal{C}$ is the set of all cliques *(a clique is defined to be any subset of sites where every pair of these sites are neighbours) and Z is the normalising constant*

$$Z = \sum_{\boldsymbol{x}} \exp\left(- \sum_{C \in \mathcal{C}} \Phi_C(\boldsymbol{x}_C) \right) < \infty. \tag{3.7}$$

The easier direction of the proof is to verify that a distribution of the form of (3.6) satisfies the Markov property (3.4). This follows from noting that

$$\pi(x_i \mid \boldsymbol{x}_{-i}) \propto \pi(\boldsymbol{x}) \propto \exp\left(- \sum_{C \in \mathcal{C}} \Phi_C(\boldsymbol{x}_C) \right). \tag{3.8}$$

Normalising the above expression over all possible values of x_i, notice that all terms $\Phi_C(\boldsymbol{x}_C)$ not involving x_i cancel out, and hence the result. It is far harder to prove the converse see, for example, Winkler (1995).

The theorem can be used in at least two ways. The first is to confirm that a collection of full conditionals (that is, the distribution of one component conditional on all the others) does define a legitimate joint density when we can find a distribution of the form (3.6) where the full conditionals match. Secondly, and more importantly, it says that instead of constructing full conditionals directly, we could construct them implicitly though the choice of so-called *potential functions* Φ_C.

3.3 Models for binary and categorical images

In this section we discuss the use of the *Ising model* and its extension, the *Potts model*, for *binary images* and images with more than two unordered colours. We will present the material in a reasonable level of detail since many of the ideas generalise quite readily to grey-level images and even high level models as well. We will begin by describing the models themselves, and their contribution to posterior distributions for images, and then discuss how to simulate from such systems.

3.3.1 Models for binary images

Suppose the image of interest is binary, where we typically refer to each pixel x_i as foreground if $x_i = 1$ (black), or background if $x_i = 0$ (white). What are our prior beliefs about such scenes $\boldsymbol{x}$? For a start, we could think of how a pixel x_i might behave conditional on everything else in the image. Assuming a four nearest neighbours scheme, consider the situation in Figure 3.3a. Here the four nearest neighbours of x_i are black, so what probability do we wish to assign to x_i also being black? Of course, this will depend on the context and what type of images we are studying, but it seems reasonable that this probability should increase with an increasing number of neighbouring black pixels. One possibility would be to use

$$\pi(x_i \text{ is black} \mid k \text{ black neighbours}) \propto \exp(\beta k) \tag{3.9}$$

where β is a positive parameter. The normalising constant here is simply $\exp(\beta$ number of white neighbours$) + \exp(\beta$ number of black neighbours$)$.

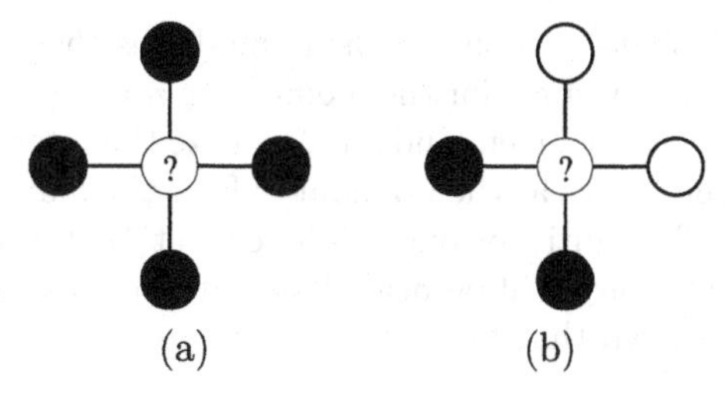

FIGURE 3.3. The pixel x_i marked as "?"and two different configurations for its four nearest neighbours

This choice implies that there is no reason to treat the background and the foreground differently, since the probabilities for $x_i = 1$ and $x_i = 0$ in Figure 3.3b are both 1/2. Also if we swap white pixels for black, and vice versa, the swapped configuration has the same probability as the original one.

We now have to appeal to the Hammersley-Clifford theorem. Can we find potential functions defined on the cliques such that the local characteristics are those we have suggested? Recall that the cliques are sets of sites such that each pair in the set are neighbours. In this case, using four nearest neighbourhoods, the cliques are the sets of nearest horizontal or nearest vertical pairs. If we try

$$\Phi_C(\boldsymbol{x}_C) = \begin{cases} -\beta, & \text{both sites in the clique } C \text{ have the same colour} \\ 0, & \text{else} \end{cases} \tag{3.10}$$

then we obtain a joint density

$$\begin{aligned} \pi(\boldsymbol{x}) &= \exp\left(-\sum_{C\in\mathcal{C}} \Phi_C(\boldsymbol{x}_C)\right) / Z(\beta) \\ &= \exp\left(\beta \sum_{i\sim j} I_{[x_i=x_j]}\right) / Z(\beta) \end{aligned} \tag{3.11}$$

where $i \sim j$ denotes neighbouring pairs. The normalising constant

$$Z(\beta) = \sum_{\boldsymbol{x}} \exp\left(\beta \sum_{i\sim j} I_{[x_i=x_j]}\right) \tag{3.12}$$

is a function of β. It is easy to verify that this has the same full conditionals as (3.9), and so our joint density is (3.11).

Obviously we did not make our full conditional suggestion at random! The joint density in (3.11) is the famous Ising model, named after E. Ising who presented the model in 1925. It has been used as a model for ferromagnetism where each site represents either an up spin or down spin. See Grimmett (1987) for a review.

One remarkable feature about the Ising model, is the existence of *phase transition* behaviour. Assume for the moment an $n \times n$ lattice, where values outside the boundary are given, and let i^* denote the interior site closest to the centre. Will the values at the boundary affect the marginal distribution of x_{i^*} as $n \to \infty$? Intuitively one might expect that the effect of what happens at the boundary will be negligible as the lattice grows, but this is wrong. It can be shown that for

$$\beta > \beta_{\text{critical}} = \log(1 + \sqrt{2}) = 0.881373\ldots \tag{3.13}$$

the effect of the boundary does matter (below this value, it does not). In consequence, long-range interaction occurs over the critical value, while only short-range interaction occurs under the critical value. We will see what this implies clearly when we simulate from the Ising model. The existence of phase transition adds weight to the assertion that it is very hard to interpret the global properties of the model by only considering the full conditionals.

3.3.2 Models for categorical images

One often encounters the situation where there are more than two colours in the image, for example in a microscopy context these might correspond to the background, cells of type 1 and cells of type 2 . One of the simplest models for such categorical settings is the Potts model, which is the multicolour generalisation of the Ising model. Suppose now that $x_i \in \{0, 1, \ldots, n_c - 1\}$, where the number of colours n_c is a positive integer greater than two, then define

$$\pi(\boldsymbol{x} \mid \beta) \propto \exp\left(\beta \sum_{i \sim j} I_{[x_i = x_j]}\right). \tag{3.14}$$

Although this expression is similar to that for the Ising model, the configurations have changed from binary to multicolour. We see from (3.14) that each of the n_c colours has the same full conditional distribution for x_i as the Ising model if we merge all neighbour sites into the two classes "same colour as x_i" and its converse. This is not unreasonable as the colours are not ordered; colour 1 is not necessary closer to colour 3 than colour 0.

Further generalisations of the Ising/Potts model

Generalisations of the Ising and Potts models can also include other neighbourhood systems than the four nearest neighbours; for example the nearest eight or twelve neighbours could be used. However, if we decide to use a larger neighbourhood system, we might also want to experiment with the potentials to promote a certain behaviour. Another natural generalisation

FIGURE 3.4. A section of a page of newsprint, displayed in reverse video.

is to allow more general β's, for example

$$\pi(\boldsymbol{x} \mid \{\beta_{\dots}\}) \propto \exp\left(\sum_{i\sim j} \beta_{ijx_i} I_{[x_i=x_j]}\right). \tag{3.15}$$

where the strength of the interaction might depend on how far in distance site i is from site j and the colour they take. However, there is always a question as to whether it is more fruitful to consider cliques of higher orders instead of continuing the theme of pairwise interactions only. See, for example, Tjelmeland & Besag (1998), where some interesting experiments along these lines are conducted.

3.3.3 Noisy images and the posterior distribution

Generally we are not able to record images exactly, observing data $\boldsymbol{y}$ rather than $\boldsymbol{x}$. Using Bayes theorem however, we know how to construct the posterior distribution of $\boldsymbol{x}|\boldsymbol{y}$, and via this we can learn about the underlying images. We will now consider two noisy versions of binary images, illustrating these two using Figure 3.4 as the true underlying image. This image is a small section of a page of newsprint; character recognition is an important imaging task. We will make the assumption that y_i and y_j $(j \neq i)$ are conditionally independent given $\boldsymbol{x}$, so that

$$\pi(\boldsymbol{y} \mid \boldsymbol{x}) = \pi(y_1 \mid x_1) \cdots \pi(y_n \mid x_n). \tag{3.16}$$

Gaussian additive noise

Suppose that the true image is degraded by additive *Gaussian noise*,

$$\boldsymbol{y} = \boldsymbol{x} + \boldsymbol{\epsilon} \tag{3.17}$$

where $\boldsymbol{\epsilon}$ is Gaussian with zero mean, zero covariance and variance σ^2, and $\boldsymbol{\epsilon}$ and $\boldsymbol{x}$ are independent. Then

$$\pi(y_i \mid x_i) \propto \frac{1}{\sigma} \exp\left(-\frac{1}{2}(y_i - x_i)^2/\sigma^2\right). \tag{3.18}$$

Note that the observed image $\boldsymbol{y}$ is recorded on a continuous scale (which may later be discretised) even though $\boldsymbol{x}$ is binary. Gaussian additive noise is quite commonly assumed as it may mimic additive noise from several sources. The posterior distribution for the true scene $\boldsymbol{x}$, is

$$\pi(\boldsymbol{x} \mid \boldsymbol{y}) \propto \exp\left(\beta \sum_{i \sim j} I_{[x_i = x_j]} - \frac{1}{2} \sum_{i=1}^{n} (y_i - x_i)^2 / \sigma^2\right) \tag{3.19}$$

$$\propto \exp\left(\beta \sum_{i \sim j} I_{[x_i = x_j]} + \sum_{i} h_i(x_i, y_i)\right) \tag{3.20}$$

where $h_i(x_i, y_i) = -\frac{1}{2\sigma^2}(y_i - x_i)^2$. Note that additive constants not depending on x_i can be removed from h_i as they will cancel in the normalising constant. The form of the posterior as given in (3.20) is generic, covering all types of conditionally independent noise by defining the h_i's as

$$h_i(x_i, y_i) = \log(\pi(y_i \mid x_i)). \tag{3.21}$$

Flip noise

Flip, or binary, noise occurs when the binary images are recorded with a Bernoulli probability model that the wrong colour is recorded. We assume that the error probability p is constant and that each pixel value is recorded independently

$$\pi(y_i \mid x_i) = \begin{cases} 1 - p & \text{if } y_i = x_i \\ p & \text{if } y_i \neq x_i \end{cases} \tag{3.22}$$

Note that $p = 1/2$ corresponds to the case where there is no information about $\boldsymbol{x}$ in $\boldsymbol{y}$. The posterior distribution can still be represented in the same way as (3.20), but with

$$h_i(x_i, y_i) = I_{[y_i = x_i]} \log(\frac{1-p}{p}), \tag{3.23}$$

where $I_{[]}$ is the indicator function.

Figure 3.5 shows some examples of degraded images using additive Gaussian noise (first column) and flip noise (second column). In the Gaussian case, we have rescaled the images for display purposes, using a linear grey scale from black to white so that 0 and 1 correspond to the minimum and maximum observed values of $\{y_i\}$. Our task might be to estimate or restore the images from such data; we will use images Figure 3.5e and Figure 3.5f in later examples.

3.3.4 Simulation from the Ising model

Models such as the Ising are sufficiently complex, despite their apparently simple structure, to require Markov chain Monte Carlo methods, see Chapter 1. However, as already mentioned, the local structure of Markov random

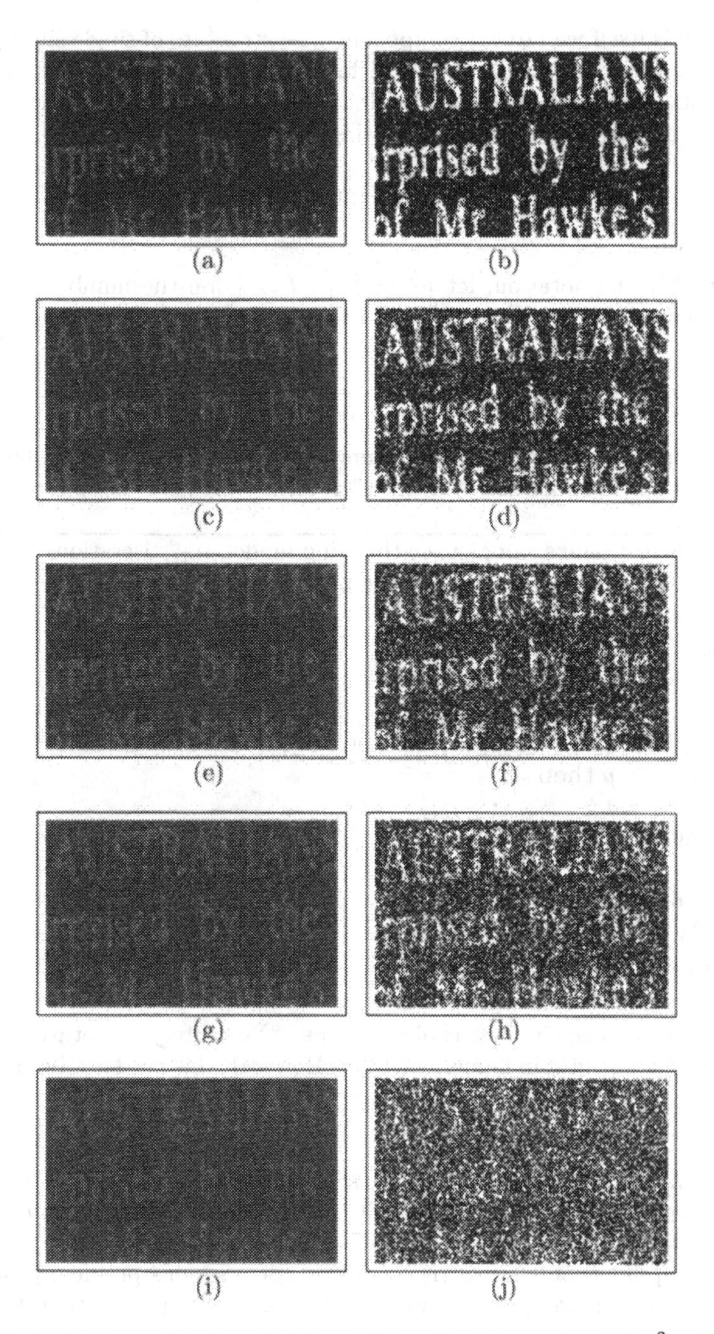

FIGURE 3.5. Newsprint degraded by additive Gaussian noise: (a) $\sigma^2 = 0.1$; (c) $\sigma^2 = 0.4$; (e) $\sigma^2 = 0.5$; (g) $\sigma^2 = 0.6$; (i) $\sigma^2 = 0.8$; or flip-noise:(b) $p = 0.1$; (d) $p = 0.2$; (f) $p = 0.25$; (h) $p = 0.3$; (j) $p = 0.4$.

fields lends itself well to the component-wise structure of single site updating versions of the family of algorithms. Recall that the *Gibbs sampler* sequentially updates each site from its full conditional. This is particularly easy as the full conditional for the Ising model is

$$\pi(x_i \mid \boldsymbol{x}_{-i}) \propto \exp(\beta \sum_{j \sim i} I_{[x_i = x_j]}) \tag{3.24}$$

To simplify the notation, let $n_i^b = \sum_{j \sim i} I_{[x_j=1]}$ be the number of black neighbours and $n_i^w = \sum_{j \sim i} I_{[x_j=0]}$ be the number of white neighbours of i, then

$$\pi(x_i = 1 \mid \boldsymbol{x}_{-i}) = \frac{\exp(\beta n_i^b)}{\exp(\beta n_i^b) + \exp(\beta n_i^w)}. \tag{3.25}$$

The implementation of the Gibbs-sampler for the Ising model is shown in Algorithm 4. The function "NextSite" returns which site to update on the

Algorithm 4 Gibbs sampler for the Ising model, n_{iter} iterations

```
Set x = 0 or 1, or fill x with randomly chosen 0's and 1's.
for t = 1 to n_iter do
  for j = 1 to n do
    i =NextSite(j)
    U ~ Uniform(0, 1)
    p = exp(βn_i^b)/(exp(βn_i^b) + exp(βn_i^w)).
    if U < p then
      x_i = 1
    else
      x_i = 0.
    end if
  end for
end for
```

j'th iteration. The simplest choice is to use "NextSite(j) = return j" that is, updating the sites in numerical order. However, this need not be the case although it is convenient from a coding point of view. Two other *updating schemes* are also commonly used:

Random With this scheme, we chose which pixel to update at random from the whole image. A benefit here is that no potential "directional effects" occur. Although there is a small chance that some pixels will be updated only a few times, this is not a serious problem provided we run our sampler for sufficiently long (the expected time to visit all the pixels is $\mathcal{O}(n \log(n))$).

Permutation With this scheme, we again chose which pixels to update at random but now with the constraint that all other pixels are updated before updating the same one again, in effect a random permutation. One implementation of this approach is to have initially a list of all sites so that each element contains a pixel index. At each call when the list is not empty, pick one element at random, return the contents and delete the element from the list. If the list is empty, then start a complete new list. There are also other approaches.

Obviously, the Gibbs proposals could be replaced equally well by a more general Hastings proposal, with the obvious modifications. One obvious choice in this binary situation is to propose to change pixel i to the other colour from its current value. If all the neighbours of i are black, the Gibbs sampler will favour x_i being black, but this will at the same time prevent the sampler from moving around, say to the symmetric configuration where all colours are reversed (and which has the same probability).

Algorithm 5 A Metropolis sampler for the Ising model

Set $\boldsymbol{x} = \mathbf{0}$ or $\mathbf{1}$, or fill $\boldsymbol{x}$ with randomly chosen 0's and 1's.
for $t = 1$ to n_{iter} **do**
  for $j = 1$ to n **do**
    $i =$NextSite(j)
    $x'_i = 1 - x_i$
    $d = \exp(\beta \sum_{j\sim i} I_{[x_i = x_j]})$
    $d' = \exp(\beta \sum_{j\sim i} I_{[x'_i = x_j]})$
    $p = \min\{1, d'/d\}$
    $U \sim$ Uniform$(0, 1)$
    if $U < p$ **then**
      $x_i = x'_i$
    end if
  end for
end for

In general, it is numerically unstable to compute the acceptance rate as in Algorithm 5. The problem arises when taking the ratio of *unnormalised* conditional densities. (Sooner or later this will give you severe problems or seemingly strange things happen with your MCMC-program, if you are not careful at this point!) A numerically more stable approach deals with the log densities for as long as possible, that is:

$d = \beta \sum_{j\sim i} I_{[x_i = x_j]}$
$d' = \beta \sum_{j\sim i} I_{[x'_i = x_j]}$
$p = \exp(\min\{0, d' - d\})$

Mixing issues

Both Algorithm 4 and 5 use single site updating, as is common in MCMC algorithms, and both perform poorly for simulating from the Ising model for β higher than or close to the critical value. Assume the current configuration is mostly black due to a high value of β. We know that the configuration formed by flipping the colours has the same probability. Using a single site sampler, we have to change the colour of all the sites individually. This involves going though a region of very low probability states, which is of course quite unlikely (although it will happen eventually), so the convergence of single site algorithms can be painfully slow for large values of β. Alternative algorithms do exist for the Ising model, most notably the *Swendsen-Wang algorithm* (Swendsen & Wang 1987) which has good convergence properties even at β_{critical}. We now describe this algorithm specifically for simulating from the Ising model, but note that more general forms exist (Besag & Green 1993); see also Section 1.3.2.

A new, high dimensional variable $\boldsymbol{u}$ is introduced, with one component of $\boldsymbol{u}$ for each interaction $i \sim j$. These u_{ij} could be thought of as bond variables. The joint distribution of $\boldsymbol{x}, \boldsymbol{u}$ is constructed by defining the conditional distribution of $\boldsymbol{u}$ given $\boldsymbol{x}$, $\pi(\boldsymbol{u}|\boldsymbol{x})$. A Markov chain is then constructed alternating between transitions on $\boldsymbol{u}$ (by drawing from $\pi(\boldsymbol{u}|\boldsymbol{x})$) and transitions on $\boldsymbol{x}$. To ensure that this two step procedure retains $\pi(\boldsymbol{x})$ as its stationary distribution, the transition function $P(\boldsymbol{x} \to \boldsymbol{x}'|\boldsymbol{u})$ is chosen to satisfy detailed balance with respect to the conditional $\pi(\boldsymbol{x}|\boldsymbol{u})$; the simplest choice for this is $P(\boldsymbol{x} \to \boldsymbol{x}'|\boldsymbol{u}) = \pi(\boldsymbol{x}'|\boldsymbol{u})$.

So, given a realisation $\boldsymbol{x}$, define the u_{ij} to be conditionally independent with

$$u_{ij} \mid \boldsymbol{x} \sim \text{Uniform}(0, \exp(\beta I_{[x_i=x_j]})). \tag{3.26}$$

That is, given $\boldsymbol{x}$, the auxiliary variable u_{ij} is uniformly distributed either on $[0, \exp(\beta)]$, if $x_i = x_j$, or on $[0, 1]$ otherwise, both of which are clearly easy to simulate. Notice that the larger β is, the more likely it is that the u_{ij} generated for a neighbouring pair $x_i = x_j$ is greater than 1.

Then, via the joint distribution of $\boldsymbol{x}$ and $\boldsymbol{u}$

$$\pi(\boldsymbol{x} \mid \boldsymbol{u}) \propto \prod_{ij} I_{[\exp(\beta I_{[x_i=x_j]}) \geq u_{ij}]}, \tag{3.27}$$

i.e. a random colouring of the pixels, subject to the series of constraints. Notice that whenever $u_{ij} \leq 1$, the constraint $\exp(\beta I_{[x_i=x_j]}) \geq u_{ij}$ is satisfied whatever the values of x_i and x_j. Conversely, if $u_{ij} > 1$, then for the constraint to be satisfied, x_i and x_j must be equal. Groups of $\boldsymbol{x}$ sites connected by some path of interactions for which each $u_{ij} > \exp(\beta I_{[x_i=x_j]})$ are known as clusters, and this definition segments $\boldsymbol{x}$ into disjoint clusters. Clusters are conditionally independent given $\boldsymbol{u}$, and can be updated separately, each to a single random colour. Notice that the larger β is, the larger the clusters are likely to be, and thus large changes can be made to $\boldsymbol{x}$ in precisely the cases where the usual algorithms struggle. It is worth remarking that, unfortunately, generalisations of the Swendsen-Wang algorithm have as yet generally lacked its spectacular success.

Some examples

We will now present some realisations from the Ising model using the Swendsen-Wang algorithm. The image is 200 × 125 (the same size used in examples later on). We have no boundary conditions, meaning that sites along the border have three neighbours, while the four corner sites have two neighbours. The examples are for a range of β values ranging from $\beta = 0.3$ in Figure 3.6a, to $\beta = 1.3$ in Figure 3.6h. Note how dramatically the image changes around β_{critical}.

It is straightforward to extend these sampling ideas to the Potts model. Figure 3.7 shows some realisations from the model on a 100 × 100 lattice with $n_c = 4$ for various β. We see the close resemblance to realisations from the Ising model, although the samples are rather more "patchy" as there are more colours present.

3.3.5 Simulation from the posterior

The posterior summarises knowledge of the true image based on our prior knowledge and the observed data. Hence, if we can provide samples from the posterior, they can be used to make inference about the true scene. We can now easily modify Algorithm 5 to account for observed data. The only change is to add the contribution from the likelihood. The convergence for this simple MCMC algorithm is, in most cases, quite good; the effect of long interactions from the prior when β is large, is reduced by the presence of the observations and so phase transition does not occur for the posterior (in most cases).

Algorithm 6 A Metropolis sampler for the noisy Ising model

Initialise $\boldsymbol{x}$
Read data $\boldsymbol{y}$ and noise-parameters
for $t = 1$ to n_{iter} **do**
  for $j = 1$ to n **do**
    $i =$NextSite(j)
    $x'_i = 1 - x_i$
    $d = \beta \sum_{j \sim i} I_{[x_i = x_j]} + h_i(x_i, y_i)$
    $d' = \beta \sum_{j \sim i} I_{[x'_i = x_j]} + h_i(x'_i, y_i)$
    $U \sim \text{Uniform}(0, 1)$
    $p = \exp(\min(d' - d, 0))$
    if $U < p$ **then**
      $x_i = x'_i$
    end if
  end for
end for

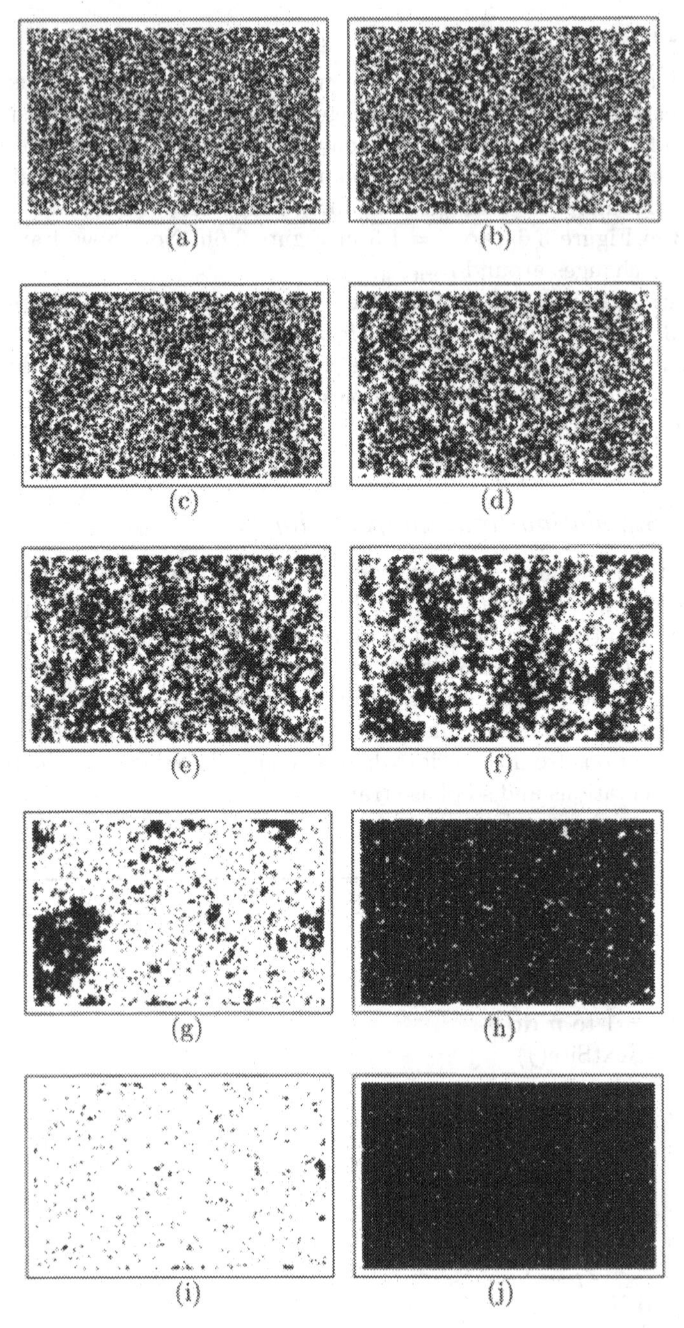

FIGURE 3.6. Simulation from the Ising model: (a) $\beta = 0.3$; (b) $\beta = 0.4$; (c) $\beta = 0.5$; (d) $\beta = 0.6$; (e) $\beta = 0.7$; (f) $\beta = 0.8$; (g) $\beta = 0.9$; (h) $\beta = 1.0$; (i) $\beta = 1.1$; (j) $\beta = 1.3$.

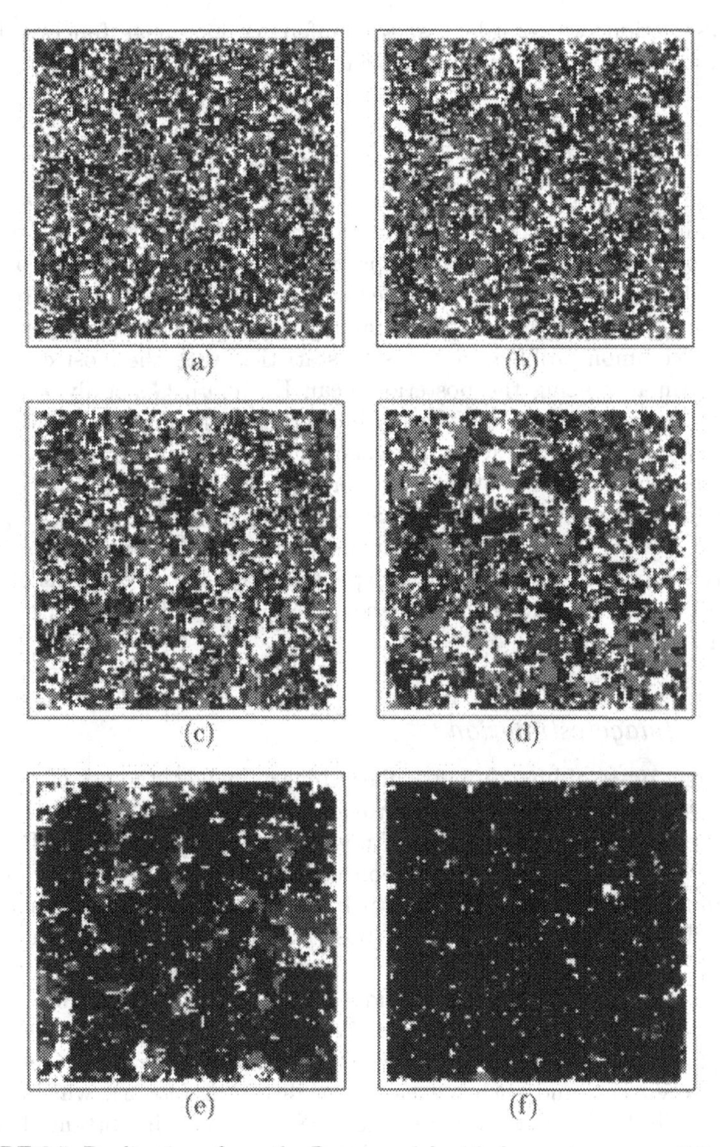

FIGURE 3.7. Realisations from the Potts-model with four colours on a 100×100 lattice: (a) $\beta = 0.7$; (b) $\beta = 0.8$; (c) $\beta = 0.9$; (d) $\beta = 1.0$; (e) $\beta = 1.1$; (f) $\beta = 1.2$.

3.4 Image estimators and the treatment of parameters

In this section, we consider issues of inference for image functionals, for images themselves and for the associated model parameters.

3.4.1 Image functionals

In some cases, it is not the image as a picture which is of primary concern, but rather some quantitative information carried by the image, for example the typical size of a cell, or the proportion of land used for growing a particular crop. It is usually possible to express such attributes as a function of $\boldsymbol{x}$, say $g(\boldsymbol{x})$. Finding posterior summaries of $g(\boldsymbol{x})$ is then recognisable as a fairly common problem in Bayesian statistics, with the most commonly used estimator being the posterior mean $\mathrm{E}_{\boldsymbol{x}|\boldsymbol{y}} g(\boldsymbol{x})$. Generally, of course, this expectation will be analytically intractable, and a run of MCMC must be used to approximate it by the usual ergodic average. This does at least have the advantage that other posterior summaries can be extracted in the course of the computation, for example credible intervals. It is perhaps this ability to get a handle on the uncertainty in imaging problems which may justify the use of computationally expensive Bayesian methods as opposed to the many other cheaper algorithms developed in the computer vision world.

3.4.2 Image estimation

For many types of images, particularly binary images, the use of posterior means as a summary does not provide sensible solutions; for that reason, we now consider the range of possibilities available. In the Bayesian framework, estimation is based upon the specification of appropriate *loss functions* (negative utility), for which we then derive the corresponding optimal estimators. Of course the posterior mean is an example of this, corresponding to the squared loss function. One interpretation of the loss function, in this context, is to consider it as a measure of distance between a true image $\boldsymbol{x}$ and an estimate $\boldsymbol{z}$. We are trying to capture the notion of how wrong we are if we mistakenly use $\boldsymbol{z}$ rather than the correct $\boldsymbol{x}$. Suppose we can find such a measure of distance, $L(\boldsymbol{x}, \boldsymbol{z})$, which defines numerically how close the two images are. It is more important that this L provides a reasonable measure of distance, rather than strictly satisfying the axioms of a metric (namely, $L(\boldsymbol{x}, \boldsymbol{z}) = L(\boldsymbol{z}, \boldsymbol{x}) \geq 0$ with equality iff $\boldsymbol{z} = \boldsymbol{x}$, and that $L(\boldsymbol{x}, \boldsymbol{z}) \leq L(\boldsymbol{x}, \boldsymbol{u}) + L(\boldsymbol{u}, \boldsymbol{z})$). Suppose we were to evaluate how close an estimate $\boldsymbol{z}$ is to the image $\boldsymbol{x}$ by using $L(\boldsymbol{x}, \boldsymbol{z})$. Different estimates could then be compared, and as an estimate $\boldsymbol{z}'$ is better than $\boldsymbol{z}''$ if $L(\boldsymbol{x}, \boldsymbol{z}') < L(\boldsymbol{x}, \boldsymbol{z}'')$. Although this is feasible when we are using a known test image $\boldsymbol{x}$, the basic concept still applies when $\boldsymbol{x}$ is unknown and we have available its posterior distribution. We can define the posterior expected distance between any estimate $\boldsymbol{z}$ and the unknown $\boldsymbol{x}$, as

$$\mathrm{E}_{\boldsymbol{x}|\boldsymbol{y}} L(\boldsymbol{x}, \boldsymbol{z}) = \sum_{\boldsymbol{x}} \pi(\boldsymbol{x} \mid \boldsymbol{y}) L(\boldsymbol{x}, \boldsymbol{z}). \tag{3.28}$$

We define the optimal Bayes estimate as the configuration minimising (3.28),

$$\widehat{\boldsymbol{x}} = \arg\min_{\boldsymbol{z}} \mathrm{E}_{\boldsymbol{x}|\boldsymbol{y}} L(\boldsymbol{x}, \boldsymbol{z}). \tag{3.29}$$

Although the general setup is straightforward, the practical implementation of these ideas is not trivial for two main reasons:

1. How should we construct a distance measure which conforms to our visual perception of how close two binary images are? Suppose there are four discrepant pixels, then the location of these pixels really matters! For example, if they are clumped, they may be misinterpreted as a feature. Ideally, a distance measure should take into account the spatial distribution of the errors as well as their number.

2. Assume we have a distance measure, how can we obtain the optimal Bayes estimate in practise, i.e. solve (3.29)?

To get some feeling as to how certain loss functions behave, consider the error vector $\boldsymbol{e}$, where $e_i = I_{[x_i \neq z_i]}$. We can expand any loss function based on a difference between images using a binary expansion

$$\begin{aligned} L(\boldsymbol{e}) &= a_0 - a_1 \sum_i (1 - e_i) - a_2 \sum_{i<j} (1 - e_i)(1 - e_j) \\ &- a_3 \sum_{i<j<k} (1 - e_i)(1 - e_j)(1 - e_k) \\ &- \ldots - a_n (1 - e_1)(1 - e_2) \cdots (1 - e_{n-1})(1 - e_n) \end{aligned} \tag{3.30}$$

where for simplicity we have assumed that the constants $(a_0, \ldots, a_n)$ depend only on the number of sites considered in each term, and the a_0 term is usually selected such that $L(\mathbf{0}) = 0$. We see that each summand is either 1 if there are no errors in the terms concerned, or 0 if there is at least one error. Two common image estimators are *marginal posterior modes* (MPM) and *maximum a posteriori* (MAP) which correspond to a loss function which counts the number of pixel misclassifications and to a loss function which is zero if there is no error and is one otherwise, respectively:

$$L_{\mathrm{MPM}}(\boldsymbol{e}) = \sum_{i=1}^{n} e_i, \tag{3.31}$$

and

$$L_{\text{MAP}}(\boldsymbol{e}) = 1 - \prod_{i=1}^{n}(1 - e_i). \tag{3.32}$$

Note that these choices corresponds to two extremes, namely that the non zero $\boldsymbol{a}$ values for MPM are $a_0 = n$, $a_1 = 1$, and for MAP $a_0 = 1$ and $a_n = 1$.

The MPM estimator

The loss function in (3.31) simply counts the number of errors, so the corresponding optimal Bayes estimate will be optimal in the sense of minimising the number of misclassifications. There is no extra penalty if the errors are clustered together as opposed to being scattered around, and in this sense the estimator is quite local. To obtain the estimate, we first compute the expected loss

$$\begin{aligned} \mathrm{E}_{\boldsymbol{x}|\boldsymbol{y}} \sum_i e_i &= \sum_i \mathrm{E}_{x_i|\boldsymbol{y}} e_i = \sum_i \Pr(x_i \neq z_i \mid \boldsymbol{y}), \\ &= \text{constant} - \sum_i \Pr(x_i = z_i \mid \boldsymbol{y}), \end{aligned} \tag{3.33}$$

hence by definition,

$$\begin{aligned} \boldsymbol{x}_{\text{MPM}} &= \arg\min_{\boldsymbol{z}} \left\{ -\sum_i \Pr(x_i = z_i \mid \boldsymbol{y}) \right) \\ &= \sum_i \arg\max_{z_i} \Pr(x_i = z_i \mid \boldsymbol{y}). \end{aligned} \tag{3.34}$$

So the ith component of $\boldsymbol{x}_{\text{MPM}}$ is the modal value of the posterior marginal. In our case, it is simply

$$x_{\text{MPM},i} = \begin{cases} 1, & \text{if } \Pr(x_i = 1 \mid \boldsymbol{y}) > 1/2 \\ 0, & \text{if } \Pr(x_i = 1 \mid \boldsymbol{y}) \leq 1/2. \end{cases} \tag{3.35}$$

To compute an estimate of $\boldsymbol{x}_{\text{MPM}}$, we can use N samples from the posterior, and if the number of times x_i is equal to 1 is greater or equal to $N/2$, then $\widehat{x}_{\text{MPM},i} = 1$, else it is 0.

The MAP estimator

The zero-one loss function (3.32) giving rise to the MAP estimate is quite extreme; any image not matching $\boldsymbol{x}$ is as wrong as any other, independent of how many errors there are. As is well known for the zero-one loss function, the estimator is

$$\boldsymbol{x}_{\text{MAP}} = \arg\max_{\boldsymbol{z}} \pi(\boldsymbol{z} \mid \boldsymbol{y}), \tag{3.36}$$

i.e. the posterior mode. The mode may be an obvious candidate for an estimator if the posterior is unimodal or has one dominating mode which also contains essentially all of the probability mass. However, this is not always the case. Much of the probability mass can be located quite far away from the mode, which makes this estimator questionable. One indication of this happening is when the mode looks quite different from "typical" realisations from the posterior.

Obviously, one way to try to find the MAP estimate would be to sample from the posterior, always keeping note of the state seen so far with the highest posterior probability. However, this is likely to be a highly inefficient strategy, and instead we next consider two alternatives, one stochastic and the other deterministic.

Algorithms for finding the MAP estimate

One algorithm which is known to converge to the MAP estimate, at least in theory, is *simulated annealing*, cf. Section 1.5.4. The basic idea is as follows: Suppose $\pi(\boldsymbol{x})$ is the distribution of interest and let $\boldsymbol{x}^*$ be the unknown mode. It would clearly be a slow and inefficient approach to search for $\boldsymbol{x}^*$ simply by sampling from $\pi(\boldsymbol{x})$, but the search could be made more efficient if we instead sample from $\pi_T(\boldsymbol{x}) \propto \pi(\boldsymbol{x})^{1/T}$ for small values of T, known as the temperature, $0 < T \ll 1$. Note that $\pi_T(\boldsymbol{x})$ has the same mode for all $0 < T < \infty$, and as $T \to 0$ will have most of its probability mass on this mode. The fact that we do not know the normalising constant as a function of temperature will not be important as we will use MCMC. So if we were to chose $T = 0^+$ and construct a MCMC algorithm to sample from $\pi_T(\boldsymbol{x})$, a sample would most likely be the mode and the problem is apparently solved! The catch is of course that the smaller T gets, the harder it is for an MCMC algorithm to mix and produce samples from $\pi_T(\boldsymbol{x})$, rather than getting trapped at a local mode of the posterior. So, the following trick is used: At iteration t, the target distribution is $\pi(\boldsymbol{x})^{1/T(t)}$, where $T(t)$ is the temperature which varies with time (hence, we have a non-homogeneous Markov chain). The temperature schedule is decreasing in such a way that $T(t) \leq T(t')$ if $t \geq t'$, and $T(t) \to 0$ as $t \to \infty$. If we decrease the temperature slowly enough, then hopefully the MCMC algorithm will reach the global mode. Theoretical analysis of the algorithm clarifies what is required of the speed of the temperature schedule. We have to lower the temperature no faster than

$$T(t) = C/\log(t+1) \tag{3.37}$$

where C is a constant depending on $\pi(\boldsymbol{x})$. Hence, the time it takes to reach $T = \epsilon$, is at least

$$t = \exp(C/\epsilon) - 1 \tag{3.38}$$

which quickly tends to infinity as ϵ tends to zero. In other words, the required schedule is not implementable in practise. Stander & Silverman

(1994) give some recommendations for schedules which perform well in finite time.

From a computational point of view, simulated annealing is easy to implement if one already has an MCMC algorithm to sample from $\pi(\boldsymbol{x})$. Note that if $\log \pi(\boldsymbol{x}) = -U(\boldsymbol{x}) + \text{constant}$, then

$$\pi(\boldsymbol{x})^{1/T(t)} \propto \exp\left(-\frac{1}{T(t)}U(\boldsymbol{x})\right), \tag{3.39}$$

so the effect of the temperature is simply a scaling of $U(\boldsymbol{x})$. In Algorithm 7 we have implemented simulated annealing using Algorithm 6. The user has to provide the "temperature" function, which returns the temperature as a function of the time. We chose n_{iter} to be finite and lower the temperature faster than (3.37), for example as

$$T(t) = T_0 \times \eta^{t-1} \tag{3.40}$$

where $T_0 = 4$ and $\eta = 0.999$. T_0, η and n_{iter} should be chosen to reach a predefined low temperature in n_{iter} iterations. Note that we may also keep track of which configuration has highest probability of those visited so far, and return that configuration as the final output.

Algorithm 7 The Simulated Annealing algorithm for the noisy Ising model

Initialise $\boldsymbol{x}$, set $T = T_0$.
Read data $\boldsymbol{y}$ and noise parameters
for $t = 1$ to n_{iter} **do**
  for $j = 1$ to n **do**
    i =NextSite(j)
    $x_i' = 1 - x_i$
    $d = \beta \sum_{j \sim i} I_{[x_i = x_j]} + h_i(x_i, y_i)$
    $d' = \beta \sum_{j \sim i} I_{[x_i' = x_j]} + h_i(x_i', y_i)$
    $U \sim$ Uniform(0, 1)
    $p = \exp(\min(d' - d, 0)/T)$
    if $U < p$ **then**
      $x_i = x_i'$
    end if
  end for
  T = Temperature(t)
end for
return $\boldsymbol{x}$

Besag (1986) introduced the method of *iterated conditional modes* (ICM) as a computationally cheaper alternative to simulated annealing. (There are also arguments for considering ICM as an estimator on its own.) It is equivalent to using simulated annealing at temperature zero taking the Gibbs sampler as the MCMC component. At each iteration, the most likely value for each pixel is chosen in turn, conditional on the current values of all the others. By considering the expression $\pi(\boldsymbol{x}|\boldsymbol{y}) = \pi(x_i|\boldsymbol{x}_{-i}, \boldsymbol{y})\pi(\boldsymbol{x}_{-i}|\boldsymbol{y})$, it is clear that each iteration increases the posterior probability until the algorithm reaches a mode, most likely a local mode. The algorithm converges fast in general, but can be very sensitive to the choice of starting point.

Finally, for the Ising model, an algorithm for locating the mode exactly exists (Greig, Porteous & Scheult 1989). The algorithm is rather technical, and not extendable beyond the Ising model, and so we do not present it here.

Examples

We will now show some estimated MPM and MAP estimates based on Figure 3.5e and f, Gaussian noise with $\sigma^2 = 0.5$ and flip noise with $p = 0.25$. We assume the noise parameter to be known, so our only *nuisance parameter* is β. It is not that common to estimate β together with $\boldsymbol{x}$, so usually the inference for $\boldsymbol{x}$ is based on

$$\pi(\boldsymbol{x} \mid \boldsymbol{y}, \beta) \tag{3.41}$$

for a fixed value of β. This is then repeated for a range of β values, and the value producing the "best" estimate is then selected. This process clearly underestimates the uncertainty regarding β, and usually also the uncertainty in other derived statistics from the posterior distribution. We will later demonstrate how this could be avoided by taking the uncertainty in β into account.

Figure 3.8 shows the case with Gaussian noise with MAP estimates in the left column and the MPM estimates in the right column. Similarly with Figure 3.9, but for the flip noise case. The effect of increasing β is clear in both sets of figures. Increasing β makes the estimate smoother. In this instance, there is not much difference between the MAP and MPM estimates.

3.4.3 Inference for nuisance parameters

We will concentrate on a binary classification model with noisy Gaussian observations of typical response levels associated with the two states (for example, our newsprint images degraded by noise). Hence, we now treat the level of background (μ_0) and foreground (μ_1) as unknown instead of being known as 0 and 1. Additionally, the noise variance σ^2 is unknown as well. The posterior then is

$$\pi(\boldsymbol{x} \mid \boldsymbol{y}) \propto \frac{(\sigma^2)^{-n/2}}{Z(\beta)} \exp\left(\beta \sum_{i \sim j} I_{[x_i = x_j]} - \frac{1}{2\sigma^2} \sum_i (y_i - \mu_{x_i})^2\right). \tag{3.42}$$

FIGURE 3.8. MAP (left column) and MPM (right column) estimates for various values of β when the true scene is degraded by Gaussian noise with variance 0.5. First row: $\beta = 0.3$, second row: $\beta = 0.5$, third row: $\beta = 0.7$, fourth row: $\beta = 0.9$, fifth row: $\beta = 1.3$.

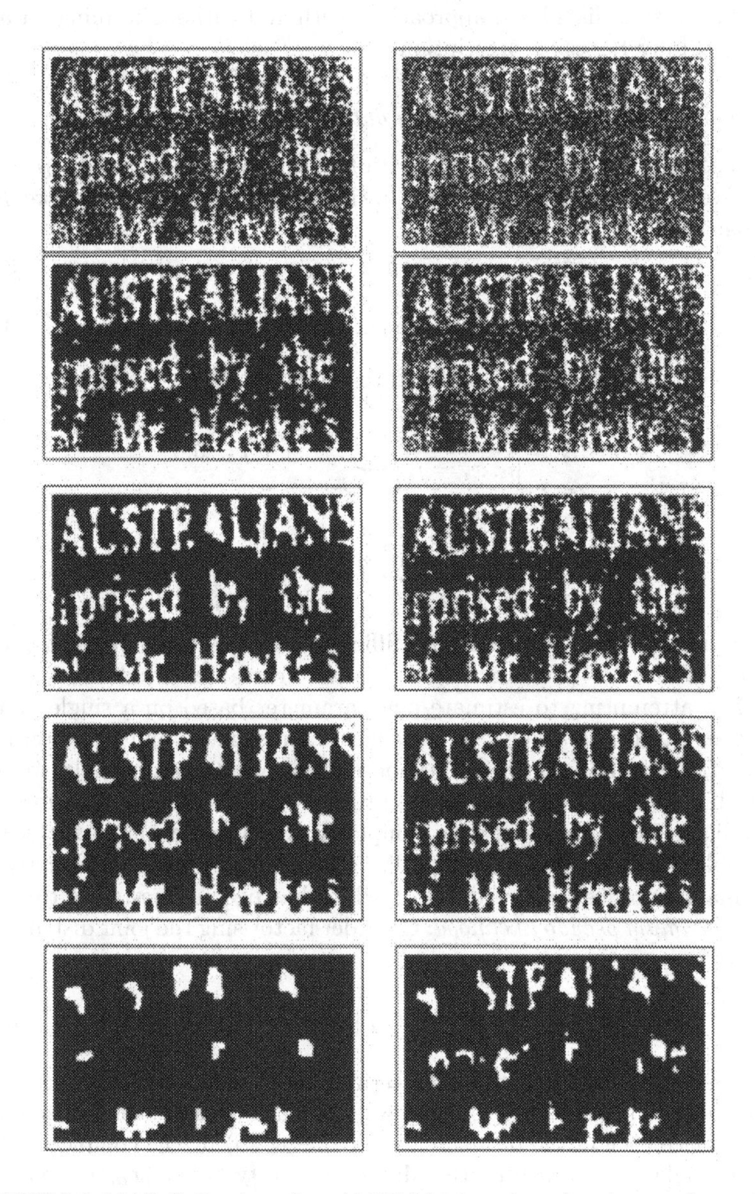

FIGURE 3.9. MAP (left column) and MPM (right column) estimates for various values of β when the true scene is degraded by flip noise with $p = 0.25$. First row: $\beta = 0.3$, second row: $\beta = 0.5$, third row: $\beta = 0.7$, fourth row: $\beta = 0.9$, fifth row: $\beta = 1.3$.

The posterior of interest has four nuisance parameters to deal with, $\beta, \sigma^2, \mu_0, \mu_1$ We will consider likelihood approaches, with and without training data, as well as a fully Bayesian description.

Likelihood approaches with training data

Suppose we are in the fortunate position where we have a data set $\boldsymbol{y}$ generated from a known configuration $\boldsymbol{x}$. In this situation, we could use *likelihood approaches*:

$$(\hat{\sigma}^2, \hat{\mu}_0, \hat{\mu}_1) = \arg\max \pi(\boldsymbol{y}|\boldsymbol{x}; \sigma^2, \mu_0, \mu_1) \tag{3.43}$$

$$\hat{\beta} = \arg\max \pi(\boldsymbol{x}; \beta). \tag{3.44}$$

For the former, it is straightforward to show that

$$\hat{\mu}_j = \frac{1}{|i : x_i = j|} \sum_{i:x_i=j} y_i, \qquad j = 0, 1 \tag{3.45}$$

$$\hat{\sigma}^2 = \frac{1}{n} \sum_i (y_i - \hat{\mu}_{x_i})^2. \tag{3.46}$$

However for $\hat{\beta}$, we have significant difficulties because the normalising constant $Z(\beta)$ for the prior distribution is not tractable. In addition, we are in effect attempting to estimate one parameter based on a single "data" point, $\boldsymbol{x}$, because although we may have many pixels, and thus considerable information about typical response levels and variance levels, we only have this single realisation from the process controlled by β. Working with high-dimensional problems with complex interaction parameters, this sort of problem arises quite frequently. We describe one possible alternative to true maximum likelihood estimation which has been suggested in this context, *maximum pseudo likelihood*: Consider factorising the joint distribution $\pi(\boldsymbol{x}|\beta)$

$$\pi(\boldsymbol{x} \mid \beta) = \pi(x_1 \mid x_2, \ldots, x_n, \beta)\pi(x_2 \mid x_3, \ldots, x_n, \beta) \ldots \pi(x_n \mid \beta) \tag{3.47}$$

Despite the Markov structure, the terms on the right-hand side above become increasingly difficult to compute. The idea behind pseudo likelihood is to replace each of the terms $\pi(x_i|x_{i+1}, \ldots, x_n, \beta)$ by the complete conditional $\pi(x_i|\boldsymbol{x}_{-i}, \beta)$ which by the Markov property is $\pi(x_i|\boldsymbol{x}_{\partial i}, \beta)$. So

$$PSL(\beta) = \prod_{i=1}^{n} \pi(x_i|\boldsymbol{x}_{\partial i}, \beta). \tag{3.48}$$

It should be noted that this is not a likelihood except in the case of no dependence, that is when $\pi(\boldsymbol{x}_i|\boldsymbol{x}_{\partial i}) = \pi(\boldsymbol{x}_i)$. The benefit of this approach is that these conditional probabilities are particularly easy to express for the Ising model:

$$\pi(x_i \mid \boldsymbol{x}_{\partial i}, \beta) = \left(1 + \exp(-\beta \sum_{j \in \partial i} (I_{[x_i = x_j]} - I_{[x_i \neq x_j]}))\right)^{-1}. \tag{3.49}$$

This function is easily calculated for a given $\boldsymbol{x}$, and so we can write down the pseudo likelihood function. Maximisation of the function over β will have to be numerical. Obviously the higher the true value of β, the worse the estimates, since we ignore the true dependence structure.

Likelihood approaches without training data

It is more typically the case that we do not have training data either for the likelihood or the prior parameters; it is then hard to implement the approaches described in the previous section exactly. One commonly used approach is to alternate iterations of whichever updating scheme is being used for $\boldsymbol{x}$ using the current parameter estimates, with steps which update the current parameter estimates treating the current $\boldsymbol{x}$ as if it were a known ground-truth. See, for example, Besag (1986).

Fully Bayesian approach

The obvious alternative to an approach which fixes the parameters at some estimated value, is to treat them as *hyperparameters* in a *fully Bayesian approach.* That is we treat $\sigma^2, \{\mu_i\}, \beta$ as variables, specify prior distributions for them and work with the full posterior

$$\begin{aligned} \pi(\boldsymbol{x}, \sigma^2, \{\mu_i\}, \beta \mid \boldsymbol{y}) &\propto \pi(\boldsymbol{y} \mid \boldsymbol{x}, \sigma^2, \{\mu_i\}, \beta)\pi(\boldsymbol{x}, \sigma^2, \{\mu_i\}, \beta) \qquad (3.50) \\ &\propto \pi(\boldsymbol{y} \mid \boldsymbol{x}, \sigma^2, \{\mu_i\})\pi(\boldsymbol{x} \mid \beta)\pi(\sigma^2)\pi(\mu_0)\pi(\mu_1)\pi(\beta) \end{aligned}$$

where we assume independent priors for each parameter. If we have no additional prior information about any of the parameters, then common choices of priors are the following (for a data set where the recorded values are in the set $\{0, \ldots, 2^8 - 1\}$, the priors for the means have been restricted to $[0, 255]$), where $1/\sigma^2$ has been reparameterised as τ

$$\begin{aligned} \tau &\sim \exp(1) \\ \mu_0 &\sim \text{Uniform}(0, 255) \\ \mu_1 &\sim \text{Uniform}(0, 255) \\ \beta &\sim \text{Uniform}(0, \beta_{max}) \text{ where say } \beta_{max} = 2. \end{aligned} \tag{3.51}$$

For the purposes of sampling, we will most likely need to know the conditional distributions of each of the parameters:

$$\begin{aligned}
\pi(\tau \mid \cdots) &\propto (\tau)^{n/2} \exp\left(-\tau(1 + 1/2 \sum_i (y_i - \mu_{x_i})^2)\right) \\
\pi(\mu_i \mid \cdots) &\propto \exp\left(-\tau/2 \sum_{j:x_j=i} (y_j - \mu_i)^2\right) I_{[0<\mu_i<255]}, \quad i = 0, 1 \\
\pi(\beta \mid \cdots) &\propto \frac{1}{Z(\beta)} \exp\left(\beta \sum_{i \sim j} I_{[x_i = x_j]}\right) I_{[0<\beta<\beta_{max}]}. \qquad (3.52)
\end{aligned}$$

The first of these conditionals is recognisable as a gamma distribution with parameters $(n/2 + 1)$ and $(1 + 1/2 \sum_i (y_i - \mu_{x_i})^2)^{-1}$, and as such would lend itself well to the Gibbs sampler. Notice that the conditional mean is quite closely related to the maximum likelihood estimate of τ. The conditionals for the two mean level parameters are recognisable as truncated Gaussians; again notice that the conditional mean is closely related to the maximum likelihood estimate. In this case, a Metropolis algorithm could be used, perhaps taking the untruncated Gaussian as the proposal density. Only the conditional for β is not recognisable as being related to a standard distribution. This suggests using some form of Metropolis-Hastings algorithm for sampling. However, once again we run into difficulties because of the intractability of the normalising constant $Z(\beta)$; in order to implement a Metropolis-Hastings algorithm here, we would need to be able to evaluate values of Z at different values of β. We now describe one approach to this problem.

Estimating $Z(\beta)/Z(\beta')$

Recall that $Z(\beta)$ is defined by

$$Z(\beta) = \sum_{\boldsymbol{x}} \exp(\beta S(\boldsymbol{x})) \qquad (3.53)$$

where $S(\boldsymbol{x})$ is the sufficient statistic $\sum_{i \sim j} I_{[x_i = x_j]}$. One possibility is to see whether the derivative of $Z(\beta)$ with respect to β is easier to estimate than $Z(\beta)$ itself.

$$\begin{aligned}
\frac{dZ(\beta)}{d\beta} &= \sum_{\boldsymbol{x}} S(\boldsymbol{x}) \exp(\beta S(\boldsymbol{x})) \\
&= Z(\beta) \sum_{\boldsymbol{x}} (S(\boldsymbol{x})) \exp(\beta S(\boldsymbol{x})) / Z(\beta) \\
&= Z(\beta)\, \mathrm{E}_{\boldsymbol{x}|\beta} S(\boldsymbol{x}). \qquad (3.54)
\end{aligned}$$

By solving this differential equation, we obtain the *path sampling* identity

$$\log\left(Z(\beta')/Z(\beta)\right) = \int_{\beta}^{\beta'} \mathrm{E}_{\boldsymbol{x}|\tilde{\beta}} S(\boldsymbol{x})\, d\tilde{\beta}, \tag{3.55}$$

see also Sections 1.5.4, 1.7.2 and 4.7.4. As we see, this trick has reduced the difficulty of the problem to one we can tackle using the following procedure:

1. Estimate

$$\mathrm{E}_{\boldsymbol{x}|\beta} S(\boldsymbol{x}) \tag{3.56}$$

for a range of various β values using posterior mean estimates based on the output from a sequence of MCMC algorithms. (These values will depend on the image size and so will need to be recalculated for each new problem, although a proper rescaling to account for (not too different) dimensions will do fine.)

2. Construct a smoothing spline $f(\beta)$ to smooth the estimated values of $\mathrm{E}_{\boldsymbol{x}|\beta} S(\boldsymbol{x})$.

3. Use numerical or analytical integration of $f(\beta)$ to compute an estimate of (3.55),

$$\log\left(\widehat{Z(\beta')/Z(\beta)}\right) = \int_{\beta}^{\beta'} f(\tilde{\beta})\, d\tilde{\beta} \tag{3.57}$$

for each pair (β, β') required.

Example

Let us now apply the fully Bayesian approach to the same examples as in Section 3.4.2. Our first task is to estimate (3.56) to compute $f(\beta)$ for evaluating the normalising constant: We ran Algorithm 5 using 3,000 iterations after burn-in, for values of β from 0 to 1.5 in steps of 0.01. We then estimated the smooth curve $f(\beta)$ using a local polynomial smoother to reduce the noise and to compute interpolated values on a fine grid, as shown in Figure 3.10. The computation of (3.57) is then trivial.

Recall that Algorithm 5, which was used to estimate Figure 3.10, has severe mixing problems for high βs due to the invariance when flipping colours. How does this affect the estimated curve in Figure 3.10? In fact very little, as the contribution to the normalising constant from each of the two modes is the same. We might expect small errors at and around the critical β, from the contribution from all the "intermediate" configurations. These may not be properly explored by Algorithm 5. However, a careful re-estimation of Figure 3.10 using the Swendsen-Wang algorithm, which does not suffer from such mixing problems, gave an indistinguishable estimate even at and around β_{critical}.

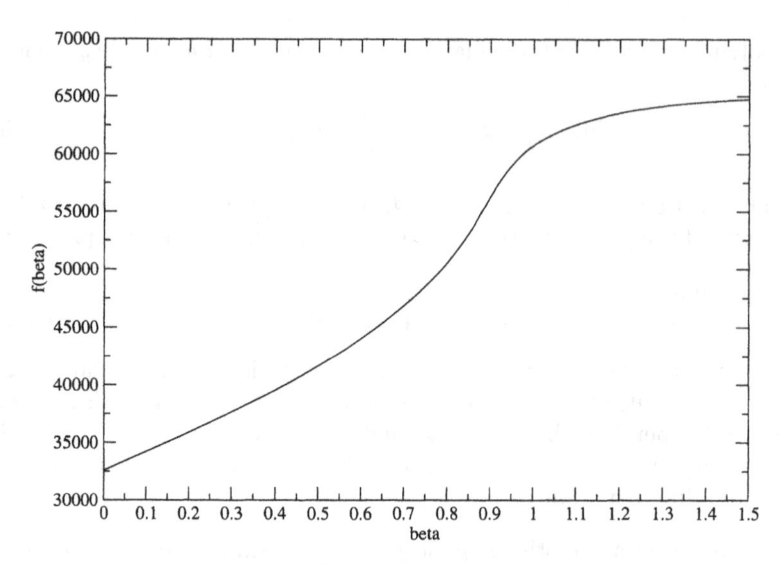

FIGURE 3.10. Estimated $f(\beta)$ for the Ising model using Algorithm 5.

We now need to extend Algorithm 6 to include a move changing β. We adopt a simple Metropolis strategy, and propose a new value β' by

$$\beta' = \beta + \text{Uniform}(-h, h) \tag{3.58}$$

where h is a small constant, in the order of $h = 0.025$ or so. The corresponding acceptance probability becomes

$$\begin{aligned} \alpha(\beta, \beta') &= \min\left\{1, \frac{\exp(\beta' S(\boldsymbol{x}) + \sum_i h_i(x_i, y_i))}{\exp(\beta S(\boldsymbol{x}) + \sum_i h_i(x_i, y_i))} \times \frac{Z(\beta)}{Z(\beta')} \times \frac{\pi(\beta')}{\pi(\beta)}\right\} \\ &= \min\left\{1, \exp\left[S(\boldsymbol{x})(\beta' - \beta) - \log\left(\frac{Z(\beta')}{Z(\beta)}\right)\right] \frac{\pi(\beta')}{\pi(\beta)}\right\} \end{aligned} \tag{3.59}$$

where $\pi(\beta)$ is the prior for β. Note that as $\beta' - \beta \to d\beta$, the exponential term in (3.59) reduces to

$$\exp\left(d\beta(S(\boldsymbol{x}) - \text{E}_{\boldsymbol{x}|\beta} S(\boldsymbol{x}))\right), \tag{3.60}$$

and so we see that β will tend to vary around the maximum likelihood estimate, $S(\boldsymbol{x}) = \text{E}_{\boldsymbol{x}|\beta} S(\boldsymbol{x})$, for β given $\boldsymbol{x}$, but with the added variability coming from $\boldsymbol{x}$ itself being unknown.

The extended sampler is then run for 2 500 iterations after burn-in for the same two noisy data sets as in Figures 3.8 and 3.9. As is clear from (3.60), the value of β is determined only by $\boldsymbol{x}$ and not by the data themselves, so by looking at Figure 3.9 and Figure 3.8, we see that the "effective" noise level is slightly higher for the flip noise case, and the estimates of the true scene are slightly more noisy. Hence, it is expected that β in the flip

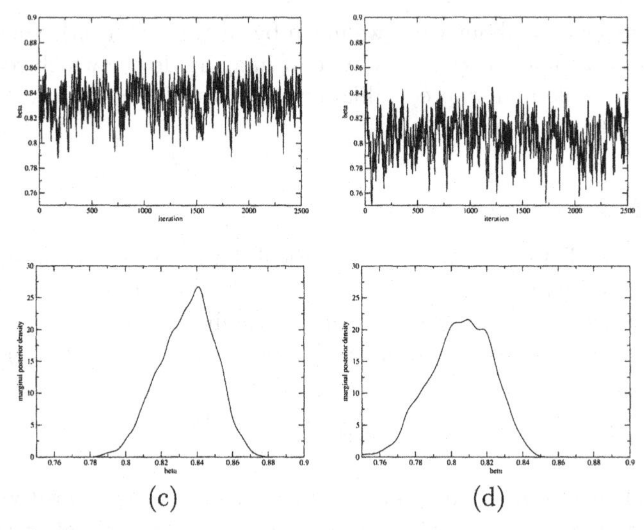

FIGURE 3.11. Figure (a) shows 2 500 values of β from the MCMC algorithm 6 and (c) the density estimate, for Gaussian noise with $\sigma^2 = 0.5$. Figures (b) and (d) are similar, but for flip noise with $p = 0.25$.

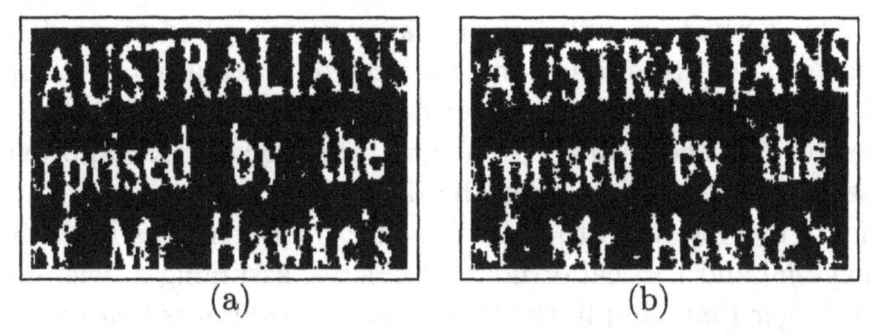

FIGURE 3.12. Posterior marginal mode estimates of the true scene with the fully Bayesian approach: (a) Gaussian noise with $\sigma^2 = 0.5$; (b) flip noise with $p = 0.25$.

noise case is slightly lower than in the Gaussian case. Figure 3.11 shows the trace plot of β and its density estimate using a kernel method, for the Gaussian case on the left, the flip noise case on the right. In both cases β varies around $0.8 - 0.85$, which is just below the critical value 0.88. The uncertainty in β might seem high for such a large data set, but in effect we have only one relevant piece of data (the whole image). Figure 3.12 shows the corresponding MPM estimates, which seem reasonable.

3.5 Grey-level images

In this section, we remain with pixel-based image models, but widen the range of application. We turn our attention to images which are *piecewise smooth*, consisting of smooth regions separated by edges (where we will be rather more precise about what we mean by smooth later on). Such images might arise in the context of any type of imaging device which records on an intensity scale and where, unlike for categorical image data, the order of these values is of interest per se.

3.5.1 Prior models

Whereas models for binary and categorical data tend to penalise any discrepancy between neighbouring pixel values, here it will be more appropriate to penalise large discrepancies more heavily than small ones. We will consider MRF prior distributions of the form $\pi(\boldsymbol{x}) \propto \exp(-\Phi(\boldsymbol{x}))$, where

$$\Phi(\boldsymbol{x}) = \beta \sum_{C \in \mathcal{C}} w_C \phi\left(D_C(\boldsymbol{x})\right), \tag{3.61}$$

ϕ is a real function, the w_Cs are positive weights, β is a positive scaling parameter, and the functions $D_C(\boldsymbol{x})$ are discrete approximations to some order of $\boldsymbol{x}$ derivative at clique C. This choice of order of derivative is one way in which to control the order of smoothness desired; penalising differences in first order derivative will favour constant regions, penalising second order derivatives will favour planar regions, and so on. The D_C's are taken to be discrete difference operator approximations, so for example an approximation to a first order derivative in the grey-level is simply the difference between values at neighbouring pixels. An approximation to the second order derivative is the difference in differences, and so on. The weights w_Cs are used to accommodate the differences in distance between diagonal and vertical or horizontal sites, assuming a neighbourhood structure larger than simply the four nearest neighbours is used.

Using the above formulation, we can write the unnormalised negative log prior as

$$\Phi(\boldsymbol{x}) = \beta \sum_{m=1}^{M} w_m \sum_i \phi\left(D_i^{(m)} \boldsymbol{x}\right), \tag{3.62}$$

where $D_i^{(m)}\boldsymbol{x}$ is the m-th discrete derivative of $\boldsymbol{x}$ at position i, e.g. $D_i^{(1)}\boldsymbol{x} = (x_{i+1} - x_i)/\delta$, where δ is a scaling parameter. The choice of the potential function ϕ has implications for the properties of the system. It is natural to assume ϕ to be symmetric, so that positive and negative gradients of equal magnitude are penalised equally, but what other properties do we desire? One particular and important possibility for ϕ is the quadratic potential $\phi(u) = u^2$ which leads to a Gaussian prior, and, combined with a Gaussian likelihood, a unimodal posterior distribution. This of course simplifies sampling; Gaussian models can be efficiently sampled using sparse matrix methods (Rue 2001), or fast Fourier transforms in the special case of homogeneous likelihood and toroidal boundary conditions (Wood & Chan 1994). However, this choice of ϕ is not suited for estimation of piecewise constant or planar fields because the rapid growth as $u \to \infty$ severely penalises the intensity jumps which may occur across edges. In addition, the slow variation around the origin might cause excessive smoothing and interpolation. Ideally we would also like to be able to detect changes between smooth regions. The choice of potential functions for edge-preserving restoration is widely discussed in the literature. We will here consider such implicit edge-preserving models; alternatives are to model the edges explicitly using discrete line processes (Geman & Geman 1984) or to use deformable templates which act on region descriptions directly (Grenander & Miller 1994).

We will largely follow Geman & Yang (1995) and consider potential functions in the class of continuous (but not necessary differentiable) functions

$$\begin{aligned} \mathcal{E} \;=\; & \Big\{\phi(\cdot) \in \mathrm{C}^{(0)}(\mathbb{R}) \mid \phi(0) = 0, \phi(u) = \phi(-u), \\ & \quad \lim_{u\to\infty} \phi(u) < \infty, \frac{d\phi}{du} \geq 0, u \in \mathbb{R}^+\Big\}. \end{aligned} \tag{3.63}$$

The arguments in favour of this class of models are largely based on heuristics, but it is clear that the finite limit and slow growth for large u ensure that intensity jumps over edges are not too severely penalised. The following example, taken from Blake & Zisserman (1987), shows that using potential functions in $\mathcal{E}$ has links to both *line processes* and robust inference.

Example 1 *Let u be a Markov random field with neighbourhood relation $\sim$, and define the line process*

$$l_{ij} = \begin{cases} 1 & \exists \text{ edge between } u_i \text{ and } u_j \\ 0 & \text{otherwise.} \end{cases}$$

Furthermore, define the negative log prior

$$\Phi(u,l) = \sum_{i\sim j} \left((u_i - u_j)^2 - 1\right)(1 - l_{ij})$$

smoothing within the disjoint regions defined by the line process l. Then Blake & Zisserman (1987) observed that

$$\begin{aligned} \inf_l \Phi(u,l) &= \sum_{i\sim j} \left((u_i - u_j)^2 - 1\right) I_{[(u_i-u_j)^2<1]} \\ &= \sum_{i\sim j} \left((u_i - u_j)^2 - 1\right)^- = \Phi^*(u), \end{aligned}$$

where $x^- = \min(x,0)$. Thus $\inf_u \inf_l \Phi(u,l) = \inf_u \Phi^*(u)$ and, in terms of modal behaviour, there is no need to model the edges explicitly, and thus instead use the truncated quadratic.

In addition to the behaviour as u grows large, the behaviour around the origin is important. Charbonnier, Blanc-Feraud, Aubert & Barlaud (1997) advocate strictly concave functions, such as

$$\phi(u) = \frac{-1}{1+|u|}, \qquad \phi(u) = \frac{|u|}{1+|u|}, \tag{3.64}$$

basing their argument on consideration of *coordinate-wise minima* (that is, a change in the value of any single pixel, results in an increase in the negative log prior). Let $\boldsymbol{x}^*$ be a coordinate-wise minimum and consider a small perturbation $x_i^* + tu$ toward the data y_i. This will lead to a decrease of order tu in the likelihood component, but an increase in the prior energy since $\phi'(0+) > 0$. By appropriately choosing the scaling parameter β the combined effect will very likely be an increase of the posterior distribution. This is in contrast to the case where $\phi'(0) = 0$, where there will be interpolation. (As an aside, note that it is generally difficult to say as much about the choice of ϕ when we consider estimators such as the posterior mean, since we then need to consider distributional properties rather than just the mode). Some potentials which have been used in the literature are given below.

Potential function	Reference
$\min(1, u^2)$	Blake & Zisserman (1987)
$u^2/(1+u^2)$	Geman & McClure (1987)
$\log\cosh u$	Green (1990)
$\log(1+u^2)$	Hebert & Leahy (1989)
$2\sqrt{1+u^2} - 2$	Charbonnier (1994)

TABLE 3.1. Some *edge preserving potentials*

3.5.2 *Likelihood models*

Generally speaking, *continuous likelihood models* for grey-level images and binary images are not so different. However, at this stage we will introduce one further aspect to the image degradation model, which is *blurring*. Blurring occurs when for some reason, which might be motion or perhaps a defect in the imaging device, spatial resolution is lost so that the value recorded at site i is actually a convolution of $\boldsymbol{x}$ values in a region around i. Denote this convolution by $\boldsymbol{z}$, then

$$z_i = (\boldsymbol{h} * \boldsymbol{x})_i = \sum_j h_j x_{i-j} \tag{3.65}$$

where the kernel $\boldsymbol{h}$ is called a *point spread function* (psf). The components of $\boldsymbol{h}$ often sum to 1, and in most cases, h_j has its largest value at $j = 0$. If

FIGURE 3.13. Examples of two types of blurring; (left) motion blur, (right) out-of-focus blur.

we have an additive noise structure, then the data $\boldsymbol{y} = \boldsymbol{z} + \boldsymbol{\epsilon}$. It is possible to show that the posterior remains a Markov random field with an enlarged neighbourhood structure.

As an example of the visual effect of blurring, Figure 3.13 simulates the effects either of constant relative motion of object and recording device or of a picture taken out of focus. The corresponding ideal point spread functions are a line of uniform weights for the motion blur, and for the out-of-focus case, radially symmetric weights defined over a pixel approximation to a circular disc.

3.5.3 Example

We end with a simulated example performing restoration of *confocal microscopy* images of human melanoma cancer cells. One such image is shown in Figure 3.14a. The true image $\boldsymbol{x}$ is degraded by blurring with a Gaussian kernel $\boldsymbol{h}$ with standard deviation of 3 pixels, and adding independent zero mean Gaussian noise with $\sigma = 15$. The resulting image $\boldsymbol{y}$ is shown in Figure 3.14b. The true image is divided into piecewise smooth regions, so it makes sense to use the edge preserving prior (3.62) for recovering the edges in the image. Obviously it should be possible to use single site MCMC algorithms here. However, because of the non-convexity of the prior and the long-range spatial interactions introduced by the point spread function $\boldsymbol{h}$, such standard samplers will converge very slowly for this model. Experience shows that updating all or parts of the variables jointly in blocks will lead to improved mixing, but for the present model, *block sampling* is only possible after reformulating the model using an idea in Geman & Yang (1995), which we present here. Introduce M auxiliary arrays $\boldsymbol{b} = (\boldsymbol{b}^{(1)}, \ldots, \boldsymbol{b}^{(M)})$, and define a distribution π^* with distribution

$$\pi^*(\boldsymbol{x}, \boldsymbol{b}) \propto \exp\left(-\beta \sum_{m=1}^{M} w_m \sum_{s \in S} \left(\frac{1}{2}\left(D_s^{(m)}\boldsymbol{x} - b_s^{(m)}\right)^2 + \psi(b_s^{(m)})\right)\right), \tag{3.66}$$

where the function $\psi(\cdot)$ is related to ϕ by the identity

$$\phi(u) = -\log \int \exp(-1/2(u-v)^2 - \psi(v))\, dv. \tag{3.67}$$

Then it is easy to show that

$$\pi(\boldsymbol{x}) = \int \pi^*(\boldsymbol{x}, \boldsymbol{b})\, d\boldsymbol{b}, \tag{3.68}$$

which means that we can use π^* to estimate the posterior mean of $\boldsymbol{x}$ under π. The motivation for this is that under the so called *dual model* π^*, $\boldsymbol{x}$ is Gaussian conditional on the data $\boldsymbol{y}$ and the auxiliary array $\boldsymbol{b}$. Let $\boldsymbol{D}^{(m)}$ be matrices representing the difference operators $\{D_s^{(m)}\}$, and let $\boldsymbol{D}^T = (\boldsymbol{D}^{(1)T}, \dots, \boldsymbol{D}^{(M)T})$. Furthermore, define $\boldsymbol{W} = \mathrm{diag}(\omega_1, \dots, \omega_M) \otimes \boldsymbol{I}_n$, and let $\boldsymbol{H}$ be a matrix representing the point spread function $\boldsymbol{h}$. Then the full conditional distribution for the true image $\boldsymbol{x}$ has distribution

$$\pi^*(\boldsymbol{x} \mid \boldsymbol{y}, \boldsymbol{b}) \propto \exp\left(-\frac{1}{2}\boldsymbol{x}^T\left(\beta \boldsymbol{D}^T\boldsymbol{W}\boldsymbol{D} + \frac{1}{\sigma^2}\boldsymbol{H}^T\boldsymbol{H}\right)\boldsymbol{x} + \boldsymbol{b}^T\boldsymbol{W}\boldsymbol{D}\boldsymbol{x}\right), \tag{3.69}$$

which is a *Gaussian Markov random field* with inverse covariance matrix $\boldsymbol{Q} = \beta\boldsymbol{D}^T\boldsymbol{W}\boldsymbol{D} + \sigma^{-2}\boldsymbol{H}^T\boldsymbol{H}$ and mean vector $\boldsymbol{\mu}^T = \boldsymbol{b}^T\boldsymbol{W}\boldsymbol{D}\,\boldsymbol{Q}^{-1}$. Assuming toroidal boundary conditions $\boldsymbol{x}$ can be sampled very efficiently using *fast Fourier transforms* as detailed in Geman & Yang (1995). In the general situation one can use *Cholesky decompositions* and *sparse matrix methods* as in Rue (2001). The components of the auxiliary array $\boldsymbol{b}$ are conditionally independent given $\boldsymbol{x}$, and can be sampled using e.g. the Metropolis-Hastings algorithm or rejection sampling.

Figure 3.14c shows a posterior mean estimate of the cell image based on 1000 iterations of the block sampler with parameters $\beta = 200$, $\delta = 50$, and $\psi(u) = |u|/(1+|u|)$. Visually the restoration is quite close to the original image, although some smoothing has taken place. The results were similar for a wide range of parameters, with smaller values of the ratio β/δ^2 leading to smoother images. Since the truth is known, a qualitative comparison between the restoration and the true image can be made. Figure 3.14d plots the *double integral distance* (Friel & Molchanov 1998) between the Markov chain samples and the true image Figure 3.14a. A constant image was used as an initial point, but the sampler seems to reach a stable state very quickly. Experiments indicate that the convergence is much faster than for the single site sampler, so the dual model formulation combined with a good block sampling algorithm seems well suited for recovering discontinuities in smooth images.

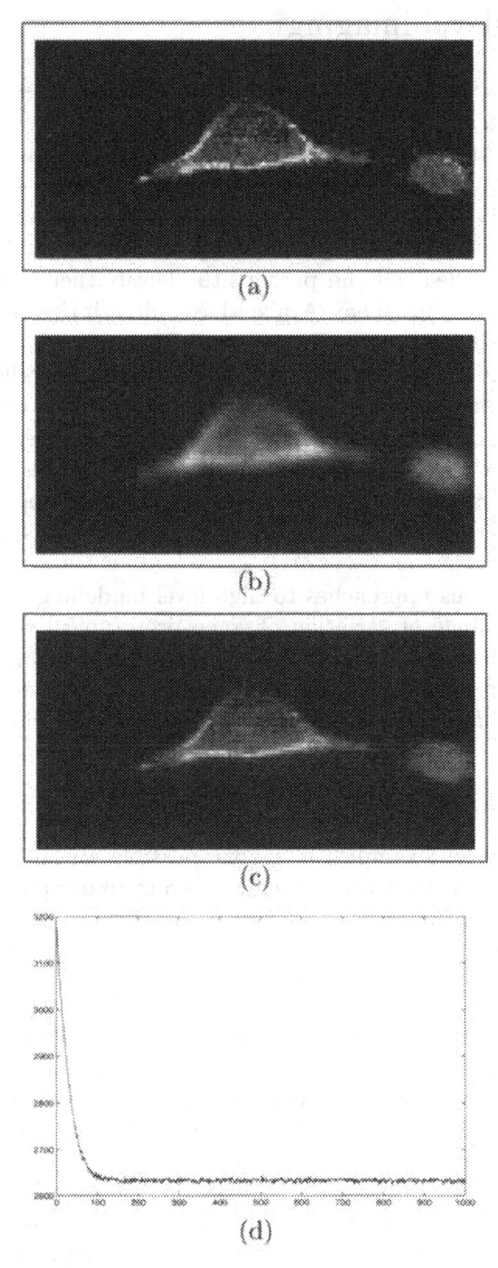

FIGURE 3.14. (a) Confocal microscopy image of a human melanoma cancer cell; (b) data simulated by blurring and adding Gaussian noise; (c) restoration of the image using the Geman & Yang model and the GMRF algorithm of Rue (2001); (d) trace plot of the double integral distance between the true image and the Markov chain samples.

3.6 High-level imaging

We now turn our attention to a completely different class of image representations and models, motivated by some of the tasks for which Markov random field representations may be inadequate. For example, suppose we are trying to identify and measure an unknown number of cells viewed under a microscope. There may be variations in the size and shape of cells, and there may even be more than one type of cell. Typical tasks might be to label the identified cell and perhaps to identify their shape characteristics, or simply to count them. A pixel-based description of the underlying $\boldsymbol{x}$ may not be the most effective one in this context for a number of reasons. Firstly, using a pixel grid, we will be "building" cells out of square building blocks. Second, if we are then trying to identify cells and we see a conglomeration of "cell" pixels, how do we know whether this is one cells or several, i.e. what constitutes a cell in terms of pixels? Third, how do we incorporate any prior information which we have about size or shape information? For problems of this sort, we may have to turn to high-level modelling.

There are various approaches to high-level modelling, mainly based on looking at the range of variation of some prototypical version of the object(s) under study (prototypical in terms of shape and/or other information such as grey-level). We will concentrate on a subset of approaches, those based on *deformations* of *polygonal templates* for objects.

3.6.1 Polygonal models

Considering our cell example above, one possible way to describe the microscope data (or at least that which is relevant to our purposes) would be to use a representation

$$\boldsymbol{x} = \{\text{Cell}_1, \text{Cell}_2, \ldots, \text{Cell}_k\} \tag{3.70}$$

where k itself may be unknown, and each Cell_i is a collection of information sufficient to locate and label that particular cell. One way to achieve this is to use a *marked point process* (see also Chapter 4) as a prior with the model for a random configuration of objects built around a stochastic template model for a single object embedded in a *mixture model* of different object types which is used as the *mark distribution* of a marked point process model. We begin, in this section, by describing the polygonal model for the outline of an object. We will describe the embedding into a point process to handle an unknown number of objects later.

The prototypical object is described by an n–sided template defined by a set of vectors $\boldsymbol{g}_0, \boldsymbol{g}_1, \ldots, \boldsymbol{g}_{n-1}$ which give the edges of the polygon, see Figure 3.15. For example, if one type of object is characteristically circular, then these edges describe a polygonal approximation to a circle. The closure

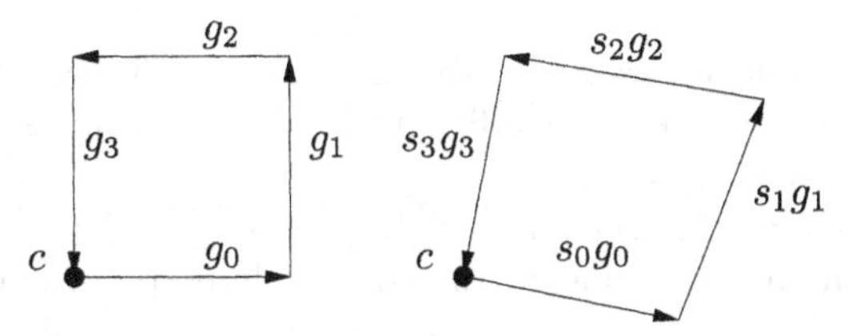

FIGURE 3.15. The left figure shows an undeformed square template. The right figures shows a deformed template holding the location c fixed.

of the polygonal template is equivalent to the condition that $\sum_{i=0}^{n-1} \boldsymbol{g}_i = \mathbf{0}$. Described in this way, the template does not have any location information, and so we will consider its first vertex to be located at the origin, the second to be located at $\boldsymbol{g}_0$, the third at $\boldsymbol{g}_0 + \boldsymbol{g}_1$ and so on. It is possible to accommodate *scaling and rotational effects*, the template may be globally scaled by a scalar R and rotated through an angle α; however we shall ignore this in this exposition. To model natural shape variability occurring between objects of the same type, each edge $\boldsymbol{g}_i$ is subject to a *stochastic deformation* which incorporates an edge–specific Gaussian deformation in length and direction. This edge-specific effect describes the change in length and direction between the undeformed $\boldsymbol{g}_i$ and the new edge. Writing the deformed edge as $\boldsymbol{s}_i\boldsymbol{g}_i$ where $\boldsymbol{s}_i$ is the 2×2 matrix representing these changes, we thus have

$$\boldsymbol{s}_i\boldsymbol{g}_i - \boldsymbol{g}_i = r_i \begin{bmatrix} \cos\theta_i & \sin\theta_i \\ -\sin\theta_i & \cos\theta_i \end{bmatrix} \boldsymbol{g}_i. \tag{3.71}$$

Writing $t_i^{(0)} = r_i \cos(\theta_i)$ and $t_i^{(1)} = r_i \sin(\theta_i)$, determines that

$$\boldsymbol{s}_i = \begin{bmatrix} 1 + t_i^{(0)} & t_i^{(1)} \\ -t_i^{(1)} & 1 + t_i^{(0)} \end{bmatrix}. \tag{3.72}$$

Specifying the distribution of r_i and θ_i to be the angular and radial components of a bivariate Gaussian with zero correlation, then $t_i^{(0)}$ and $t_i^{(1)}$ are independent Gaussians with mean zero. Ignoring for a moment the constraint that the deformed template must be closed, i.e. $\sum_{i=0}^{n-1} \boldsymbol{s}_i\boldsymbol{g}_i = \mathbf{0}$, Grenander, Chow & Keenan (1991) suggest a *first order cyclic Markov structure* on the $\{t_i^{(0)}\}$ and the $\{t_i^{(1)}\}$ independently with each having an n–dimensional Gaussian distribution with mean $\mathbf{0}$ and circulant inverse covariance matrix

$$\boldsymbol{\Sigma}^{-1} = \begin{bmatrix} \beta & \delta & & & & \delta \\ \delta & \beta & \delta & & & \\ & \delta & \beta & \delta & & \\ & & \ddots & \ddots & \ddots & \\ & & & \delta & \beta & \delta \\ \delta & & & & \delta & \beta \end{bmatrix}, \tag{3.73}$$

where all other entries are zero. We will in Section 3.7 use a second order model in (3.90), but we use the first order model in this section to avoid unnecessary details.

Define the length $2n$ vector $\boldsymbol{t} = (t_0^{(0)}, t_1^{(0)}, \ldots, t_{n-1}^{(0)}, t_0^{(1)}, \ldots, t_{n-1}^{(1)})^T$ then, still considering only the unconstrained non-closed polygon case

$$\boldsymbol{t} \sim N_{2n}\left(\mathbf{0}, \begin{bmatrix} \boldsymbol{\Sigma} & 0 \\ 0 & \boldsymbol{\Sigma} \end{bmatrix}\right). \tag{3.74}$$

Imposing the closure constraint on the deformed template will destroy the simple structure of (3.74). However, for the purposes of simulation, which will of course require MCMC methods, the unconstrained density suffices because in the acceptance ratio all we will need to evaluate is the ratio of the constrained density at two values of $\boldsymbol{x}$. Since the closure constraint is linear in $\boldsymbol{x}$, the ratio of the constrained densities is also the ratio of the unconstrained densities at the same $\boldsymbol{x}$ values.

3.6.2 Simulation

Simulation from the prior is straight forward as the joint density for the deformations are joint Gaussian and so also with the constrained density as the constraints are linear. Figure 3.16 shows some realisations from the second order model in (3.90) which we will use later in Section 3.7, for various parameter-setting. As we see, the samples mimic quite well circular-like objects.

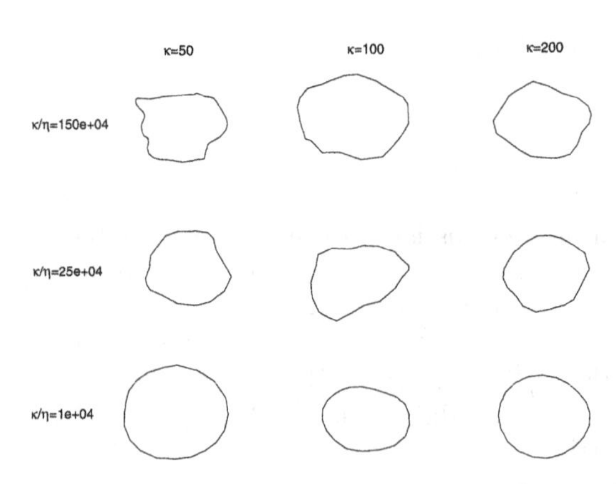

FIGURE 3.16. Samples for the edge transformation template model with precision matrix (3.90), and with different values of the parameters κ and η.

Simulation from the posterior usually also requires an MCMC approach. Consider the square template shown in Figure 3.17c. We will use this template in locating the object observed in the data shown in Figure 3.17b.

These data have been generated from the underlying pixellated image $\boldsymbol{x}$ in Figure 3.17a by setting the black background pixels to have value $\mu_0 = 0$ and the white foreground pixels to have value $\mu_1 = 1$. Pixel-wise independent zero mean Gaussian noise with variance $\sigma^2 = 0.5^2$ has then been added. Using a uniform prior density for location c, and the model for the deformations $\boldsymbol{s}$ described previously conditioned on the closure of the polygon, together with the likelihood $L(\boldsymbol{y} \,|\, \boldsymbol{s}, c)$, the posterior density for c and $\boldsymbol{s}$ becomes

$$\pi(\boldsymbol{s}, c \mid \boldsymbol{y}) \propto \pi(\boldsymbol{s}) \prod_{i \in \mathcal{I}} \exp\left(-\frac{1}{2\sigma^2}(y_i - \mu(i;\, \boldsymbol{s}, c))^2\right), \tag{3.75}$$

where $\mu(i;\, \boldsymbol{s}, c)$ is equal to μ_1 if pixel i is inside the deformed template, and equal to μ_0 if it is outside.

There are various ways in which we can propose a change to the current realisation of $\boldsymbol{x} = \{c, \boldsymbol{s}\}$. Proposing a change in c alters the location of the current polygon without altering its shape; in the early stages if the object is not well located, moves of this type are quite useful. By proposing symmetric changes in c, the acceptance ratio will depend only on the likelihood ratio of the new and old states. How can the shape of the polygon be changed? Proposed changes can be made either by altering $\boldsymbol{s}$ itself, or by altering the position of one or more of the vertices directly. Notice that the vertex locations can be written as a one-to-one linear transformation of c and $\boldsymbol{s}$ (subject to keeping the labelling of the vertices constant). This means we can propose to move a randomly chosen vertex, perhaps uniformly in some small disc around its existing value (a symmetric proposal). Because the transformation from c and $\boldsymbol{s}$ to the vertices is linear, the Jacobian is a constant, which will therefore cancel in the acceptance ratio. This allows a fast evaluation of the density $\pi(\boldsymbol{s})$ by using its first order Markov property. Figure 3.17d shows some samples of the posterior density using the template in Figure 3.17c.

There are two additional tasks that require our attention working with polygon models, that is to check if the deformed polygon is simple, and whether a site is inside or outside the deformed polygon. These tasks are classical problems in *computational geometry* , see e.g. O'Rourke (1998) for solutions and algorithms.

3.6.3 Marked point process priors

Within this framework, each object x_i comprises a point which gives location (unspecified by the deformation model) and a set of marks which then specify its other attributes, in this case its outline (Baddeley & Van Lieshout 1993). The points lie in a window Λ related to the data observation coordinates, and the marks in the space $\mathcal{M}$ associated with the object shapes. A configuration of objects is described as a finite unordered set

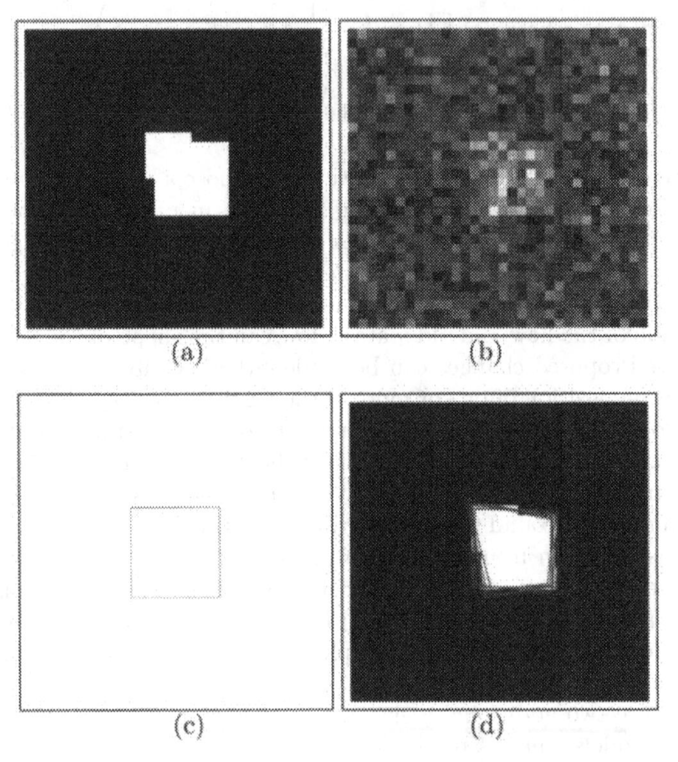

FIGURE 3.17. (a) The true pixellated image; (b) the data; (c) the square template; (d) samples of the deformed template overlaid on the true scene.

$\boldsymbol{x} = \{x_1, \ldots, x_k\}$ where $\boldsymbol{x}$ follows an object process, i.e. a marked point process on $\Lambda \times \mathcal{M}$ with a *Poisson object process* as the basic reference process (see Chapter 4). Under the Poisson process, conditional on the number of objects k, the objects are uniformly and identically distributed. The joint density of $\boldsymbol{x} = \{x_1, \ldots, x_k\}$ is defined by a density $f(\boldsymbol{x})$ relative to the Poisson object process. For example, to model *pairwise interactions* between objects which are defined to be neighbours by some relation $\sim$, the function

$$f(\boldsymbol{x}) \propto \gamma^k \prod_{i \sim j} h(x_i, x_j), \quad \gamma > 0, \tag{3.76}$$

could be used; Van Lieshout (1995) discusses various interaction models. To model a situation where objects are not allowed to overlap, such as confocal microscopy where the images are optical sections, all objects are defined to be neighbours and the interaction function $h(x_i, x_j)$ is taken to be zero if objects x_i and x_j overlap and one otherwise. This model is known as the *hard core object process*. We take the point, denoted $\boldsymbol{c}$, to be the location of the first vertex for each object. The marks are the deformations $\boldsymbol{t}$ of the standard template. It is possible to allow objects of different types by forming a mixture model using different basic templates with different deformation parameters, where the weights in the mixture model represent the relative frequencies of occurrence; refer to Rue & Hurn (1999) for details. This mixture distribution is then used as the mark distribution of the marked point process model to model an unknown number of objects.

Extending the MCMC algorithm to handle an unknown number of objects requires modifications to accommodate the dimensionality changes. Essentially, as well as proposing fixed dimension changes to the current cells, the method has to allow for *births and deaths* of cells. A framework has been provided for this by Geyer & Møller (1994), Green (1995) and Geyer (1999); see Section 4.7.7. One way to decrease the number of cells by one is to delete a randomly selected cell. The complimentary move, increasing the number of cells by one, is to propose a new cell by simulating a location uniformly in the window, and a set of marks at random from the mixture model of template types. In these move types, it is necessary to "*dimension match*" $\boldsymbol{x}$ and $\boldsymbol{x}'$, including an additional Jacobian term in the acceptance ratio. Both types of move are required in sampling our posterior density; moves which keep the dimension of $\boldsymbol{x}$ fixed, for example altering the location of shape of one of the cells, and moves which alter the dimension by altering the number of cells.

3.6.4 Parameter estimation

There are clearly some parameters in the deformable template set-up which are hard to interpret intuitively, in particular β and δ of (3.73). This is a barrier to their more widespread use in applied work. It would be nice to be able to treat them in a fully Bayesian manner, so that uncertainty could be propagated through. Unfortunately, the normalising constants of this type of model are extremely complex, and this is likely to preclude this approach. Instead in this section we will consider maximum likelihood estimation of β and δ based on recorded vertex information (as could, for example, be gathered using the mouse on a display of training data).

We begin by considering a single object, and transforming from the deformation model for the polygon edges to the model for the corresponding vertex locations. Recall that the first vertex defines the location of the entire polygon; if the first vertex is at location $\boldsymbol{c} = \boldsymbol{v}_0$, then the second vertex $\boldsymbol{v}_1$ is located at $\boldsymbol{v}_0 + \boldsymbol{s}_0\boldsymbol{g}_0$ and so on,

$$\boldsymbol{v}_j = \boldsymbol{v}_0 + \sum_{i=0}^{j} \boldsymbol{s}_i\boldsymbol{g}_i, \qquad j = 1, \ldots, n.$$

There are $n+1$ vertices in the non-closed polygon. Considering the x and y components separately, the vertices can be written

$$\begin{bmatrix} v_1^x \\ v_2^x \\ \vdots \\ v_n^x \\ v_1^y \\ v_2^y \\ \vdots \\ v_n^y \end{bmatrix} = \begin{bmatrix} g_0^x & & & & g_0^y & & & \\ g_0^x & g_1^x & & & g_0^y & g_1^y & & \\ \vdots & \vdots & \ddots & & \vdots & \vdots & \ddots & \\ g_0^x & g_1^x & \cdots & g_{n-1}^x & g_0^y & g_1^y & \cdots & g_{n-1}^y \\ g_0^y & & & & -g_0^x & & & \\ g_0^y & g_1^y & & & -g_0^x & -g_1^x & & \\ \vdots & \vdots & \ddots & & \vdots & \vdots & \ddots & \\ g_0^y & g_1^y & \cdots & g_{n-1}^y & -g_0^x & -g_1^x & \cdots & -g_{n-1}^x \end{bmatrix} \begin{bmatrix} t_0^{(0)} \\ t_1^{(0)} \\ \vdots \\ t_{n-1}^{(0)} \\ t_0^{(1)} \\ t_1^{(1)} \\ \vdots \\ t_{n-1}^{(1)} \end{bmatrix} + \bar{\boldsymbol{v}}_0 \tag{3.78}$$

where $\bar{\boldsymbol{v}}_0$ is the vector of vertex x and y positions of the undeformed template with first vertex located at $\boldsymbol{v}_0$. We will write (3.78) in the form $\boldsymbol{v} = \boldsymbol{G}\boldsymbol{t} + \bar{\boldsymbol{v}}_0$. The distribution of $\boldsymbol{v}$ unconstrained by closure given the observed $\boldsymbol{v}_0$ is therefore

$$\boldsymbol{v}^T \mid \boldsymbol{v}_0 \sim N_{2n}\left(\bar{\boldsymbol{v}}_0 \, , \, \boldsymbol{G} \begin{bmatrix} \boldsymbol{\Sigma} & \boldsymbol{0} \\ \boldsymbol{0} & \boldsymbol{\Sigma} \end{bmatrix} \boldsymbol{G}^T \right). \tag{3.79}$$

To find the constrained distribution of $(\boldsymbol{v}_1, \ldots, \boldsymbol{v}_{n-1})^T | (\boldsymbol{v}_n = \boldsymbol{v}_0)$, it is simpler to reorder the components of $\boldsymbol{v}$ from x then y components to the vertex pairs, rewriting (3.79) as

$$(\boldsymbol{v}_1, \boldsymbol{v}_2, \ldots \boldsymbol{v}_n)^T \mid \boldsymbol{v}_0 \sim N_{2n}\left(\begin{bmatrix} \boldsymbol{\mu}_1 \\ \boldsymbol{v}_0 \end{bmatrix}, \begin{bmatrix} \boldsymbol{\Sigma}_{11} & \boldsymbol{\Sigma}_{12} \\ \boldsymbol{\Sigma}_{12}^T & \boldsymbol{\Sigma}_{22} \end{bmatrix} \right), \tag{3.80}$$

where the partitioning of the mean and variance correspond to (3.79) partitioned into the sets $(\boldsymbol{v}_1, \ldots, \boldsymbol{v}_{n-1})^T$ and $\boldsymbol{v}_n^T$. Note that $\mathrm{E}(\boldsymbol{v}_n | \boldsymbol{v}_0) = \boldsymbol{v}_0$ by closure of the undeformed template. Denote the partitioned inverse of the variance matrix

$$\begin{bmatrix} \boldsymbol{\Sigma}_{11} & \boldsymbol{\Sigma}_{12} \\ \boldsymbol{\Sigma}_{12}^T & \boldsymbol{\Sigma}_{22} \end{bmatrix}^{-1} = \begin{bmatrix} \boldsymbol{\Psi}_{11} & \boldsymbol{\Psi}_{12} \\ \boldsymbol{\Psi}_{12}^T & \boldsymbol{\Psi}_{22} \end{bmatrix}. \tag{3.81}$$

Denote $\boldsymbol{v}_1, \ldots, \boldsymbol{v}_{n-1}$ by $\boldsymbol{v}_{-n}$, then

$$\pi(\boldsymbol{v}_{-n}, \boldsymbol{v}_n \mid \boldsymbol{v}_0) = \frac{(2\pi)^{-n}}{|\boldsymbol{\Sigma}|^{1/2}} \exp\left(-\frac{1}{2} \begin{bmatrix} \boldsymbol{v}_{-n} - \boldsymbol{\mu}_1 \\ \boldsymbol{v}_n - \boldsymbol{v}_0 \end{bmatrix}^T \boldsymbol{\Sigma}^{-1} \begin{bmatrix} \boldsymbol{v}_{-n} - \boldsymbol{\mu}_1 \\ \boldsymbol{v}_n - \boldsymbol{v}_0 \end{bmatrix} \right) \tag{3.82}$$

$$\pi(\boldsymbol{v}_n \mid \boldsymbol{v}_0) = \frac{(2\pi)^{-1}}{|\boldsymbol{\Sigma}_{22}|^{1/2}} \exp\left(-\frac{1}{2} \left[\boldsymbol{v}_n - \boldsymbol{v}_0\right]^T \boldsymbol{\Sigma}_{22}^{-1} \left[\boldsymbol{v}_n - \boldsymbol{v}_0\right] \right). \tag{3.83}$$

Then the conditional density of interest is

$$\begin{aligned} \frac{\pi(\boldsymbol{v}_{-n}, \boldsymbol{v}_n \mid \boldsymbol{v}_0)|_{\boldsymbol{v}_n = \boldsymbol{v}_0}}{\pi(\boldsymbol{v}_n \mid \boldsymbol{v}_0)|_{\boldsymbol{v}_n = \boldsymbol{v}_0}} &= (2\pi)^{-(n-1)} \left(\frac{|\boldsymbol{\Sigma}|}{|\boldsymbol{\Sigma}_{22}|} \right)^{-1/2} \\ &\times \exp\left(-\frac{1}{2} (\boldsymbol{v}_{-n} - \boldsymbol{\mu}_1)^T \boldsymbol{\Psi}_{11} (\boldsymbol{v}_{-n} - \boldsymbol{\mu}_1) \right). \end{aligned} \tag{3.84}$$

By some manipulation of the variance matrices, this can be seen to be the density of a $N_{2n-2}(\boldsymbol{\mu}_1, \boldsymbol{\Psi}_{11}^{-1})$. Assuming that $\boldsymbol{v}_0$ is uniformly distributed in the observation window, and under an assumption of independence of the polygon shapes, the likelihood for m cells will be the product of these individual likelihoods. This will have to be maximised numerically.

3.7 An example in ultrasound imaging

In this final section, we present an analysis of a real data set. Our goal here is to demonstrate how complex tasks can be tackled using techniques based on the type of ideas presented in the previous sections.

3.7.1 Ultrasound imaging

Ultrasound is widely used in medical settings, mainly because of its ease of use and its real-time imaging capability. However, the diagnostic quality of ultrasound images is low due to noise and image artifacts (*speckle*) introduced in the imaging process. The principle of ultrasound imaging is simple: A pulse of ultra-high frequency sound is sent into the body, and the backscattered return signal is measured after a time delay corresponding to depth. When the pulse hits a boundary between tissues having different acoustic impedances, it is partially reflected and partially transmitted. In addition there is reflection within homogeneous material due to small spatial variations in acoustical impedance, called scatterers. Thus variations in acoustic impedance is the basis for identifying regions of interest in the imaged tissue. We concentrate on the second mode of variation, called diffuse scattering, and use a Bayesian model developed in Hokland & Kelly (1996) and Husby, Lie, Langø, Hokland & Rue (2001). In this model the area Ω_i imaged in pixel i is assumed to consist of a large number of uniformly distributed scatterers, and the received signal is the sum of the reflections from all these scatterers. Assuming that all scatterers are independent, and invoking the central limit theorem, the resulting radio frequency signal x_i is assumed to be a Gaussian random variable with mean zero and a variance determined by the scattering properties of the tissue in Ω_i,

$$x_i \mid \sigma_i^2 \sim \mathrm{N}\left(0, \sigma_i^2\right), \forall i. \tag{3.85}$$

In other words the variance characterises the different tissues, and we can thus segment the image into different tissue types by identifying regions where the Gaussian echoes have approximately equal variances. Note that given the variances, the radio-frequency signal contains no additional information, and can thus be regarded as a nuisance parameter.

The observed image $\boldsymbol{y}$ is modelled as resulting from a convolution of the radio frequency signal $\boldsymbol{x}$ with the imaging system point spread function $\boldsymbol{h}$, with the addition of independent and identically distributed Gaussian noise. We assume the point spread function to be spatially invariant, thus

$$y_i \mid \boldsymbol{x} \sim \mathrm{N}\left(\sum_k h_k x_{i+k}, \tau^2\right), \quad \forall i, \tag{3.86}$$

where τ^2 is the noise variance. The pulse function is modelled as a separable Gaussian function with a sine oscillation in the radial direction, i.e.

$$h_{k,l} \propto \exp\left(-\frac{k^2}{2\sigma_r^2} - \frac{l^2}{2\sigma_l}\right) \cos\frac{2\pi k}{\omega}. \tag{3.87}$$

Empirical studies indicate that this is a good approximation which seems to be quite robust with respect to misspecification of the parameters.

Figure 3.18 shows examples of medical ultrasound images. The images are log-compressed before display to make it easier to see the anatomy. Figure 3.18a shows a part of the right ventricle of a human heart, while Figure 3.18b shows a cross-section of the carotid artery. Figure 3.18c shows an abdominal aorta aneurism, that is a blood-filled dilatation of the aorta. In the middle of the aorta it is possible to spot a vessel prothesis. Common to the images is the characteristic speckle pattern that makes it difficult for the untrained eye to spot the important anatomical features. We will focus on the cross-sectional images of the carotid artery, doing both image restoration and estimation of the cross-sectional area. An interval estimate of the artery area might also be useful as a mean of diagnosing atherosclerosis.

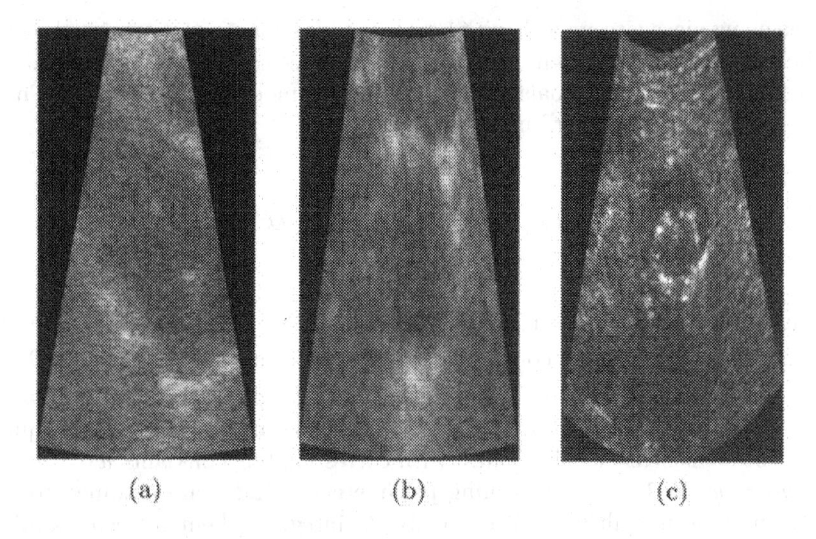

FIGURE 3.18. Real ultrasound images. Panel (a) and (b) show log-compressed radio frequency images of (a) the right ventricle of the heart, and (b) a cross section through the carotid artery. Panel (c) shows a B-scan ultrasound image of an aorta aneurism with a vessel prothesis in the middle.

3.7.2 Image restoration

We first consider restoration of the true radio-frequency image given the observed image $\boldsymbol{y}$. In this respect the most important modelling step is the specification of a prior model for the underlying variance field $\boldsymbol{\sigma}^2$, since this parameter contains information about the anatomy of the imaged tissue. In fact, the radio frequency field $\boldsymbol{x}$ contains no additional information, and can thus be seen as a nuisance parameter. However, as argued in Husby et al. (2001), a model formulation containing $\boldsymbol{x}$ has great computational advantages, since the distribution for the variance field $\boldsymbol{\sigma}^2$ given the data has no local Markov structure, whereas the distribution for $\boldsymbol{\sigma}^2$ given the data and the radio frequency field has a neighbourhood structure depending on the support of the point spread function $\boldsymbol{h}$.

To avoid problems with positivity we reparameterise the model and define a log-variance field $\boldsymbol{\nu} = (\log \sigma_i \, : \, i \in \mathcal{I})$. The choice of prior model for this field should be justified from physical considerations about the imaged tissue, and we use the following assumptions:

- the scattering intensity tends to be approximately constant within regions of homogeneous tissue,
- abrupt changes in scattering intensity may occur at interfaces between different tissue types.

Based on these assumptions it is reasonable to model the log-variance field

$\boldsymbol{\nu}$ as being piecewise smooth with homogeneous subregions corresponding to the different tissue types in the imaged region. As explained in Section 3.5.1, edge-preserving functionals are well suited for modelling such fields. Thus we define the prior distribution for $\boldsymbol{\nu}$ as

$$\pi(\boldsymbol{\nu}) \propto \exp\left(-\beta \sum_{m=1}^{M} w_m \sum_{i \in \mathcal{I}} \phi\left(D_i^{(m)}\boldsymbol{\nu}\right)\right), \tag{3.88}$$

where ϕ is a functional from the edge preserving class defined in (3.62), $w_1, \ldots, w_M$ are positive constants, β is a positive scaling factor, and $D_i^{(m)}$ are difference operators approximating first order derivatives, e.g. $D_i^{(1)}\boldsymbol{\nu} = (\nu_{i+1} - \nu_i)/\delta$. Unless otherwise stated, we will use the four first order cliques (and so eight nearest neighbours) with corresponding constants $w_1 = w_2 = 1$, $w_3 = w_4 = 1/\sqrt{2}$. The scaling parameters β and δ are assumed to be known constants, although it is possible to integrate them out using a fully Bayesian approach. δ should be selected to match the intensity jumps in the image, while a suitable choice for β can be found by trial and error. Large values of β tend to give smooth realisations, while small values give noisy realisations.

Combining equations (3.85), (3.86) and (3.88) we obtain the full conditional distribution for the radio-frequency image $\boldsymbol{x}$ and the log-variance field $\boldsymbol{\nu}$ as

$$\begin{aligned} \pi(\boldsymbol{x}, \boldsymbol{\nu} \mid \ldots) \quad \propto \quad & \prod_{i \in \mathcal{I}} \exp\left(-\frac{1}{2\tau^2}\left(y_i - \sum_k h_k x_{i+k}\right)^2\right) \qquad (3.89) \\ \times \quad & \exp\left(-\frac{x_i^2}{2}\exp(-2\nu_i) - \nu_i - \beta \sum_m \omega_m \phi\left(D_i^{(m)}\boldsymbol{\nu}\right)\right). \end{aligned}$$

A point estimate $\hat{\boldsymbol{x}}$ of the true radio frequency image $\boldsymbol{x}^*$ can be constructed using MCMC output, and a natural first choice of estimator is the posterior mean. The simplest way of constructing the Markov chain is to use a single-site Metropolis-Hastings algorithm for $\boldsymbol{\nu}$, and the Gibbs sampler for $\boldsymbol{x}$. Alternatively, one might update $\boldsymbol{\nu}$ and $\boldsymbol{x}$ as blocks by using the same dual formulation as before and utilising an algorithm for efficient sampling of Gaussian Markov random fields (Rue 2001). The single site sampler was run for 10,000 iterations on the blood vessel image in Figure 3.19a, and posterior mean estimates of the radio frequency- and log-variance-fields are shown in Figure 3.19b and c, respectively. The images are plotted in polar coordinates. To get a feel for the convergence of the chain, we have plotted traces of the log-variance at two different positions (Figure 3.20a and b), as well as the functional $f(\boldsymbol{\nu}) = \beta \sum_m \omega_m \sum_i \phi(D_i^{(m)}\boldsymbol{\nu})$ (Figure 3.20c).

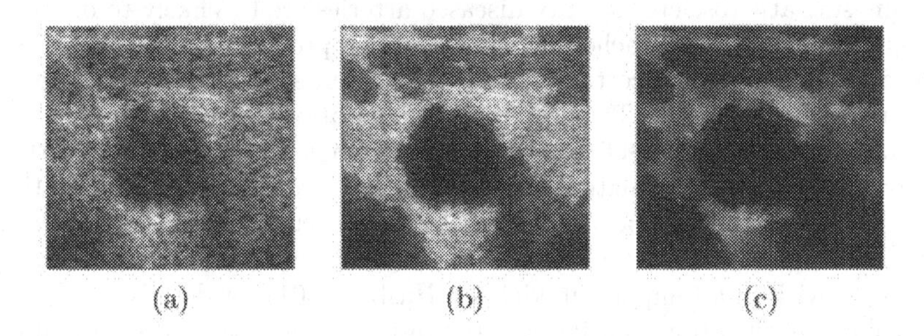

FIGURE 3.19. Image restoration using an edge-preserving model: (a) log-compressed radio frequency image of a cross section through the carotid arteryl in polar coordinates; (b) a posteriori mean estimate of the true radio frequency image; (c) the corresponding posterior mean estimate of the underlying log-variance field.

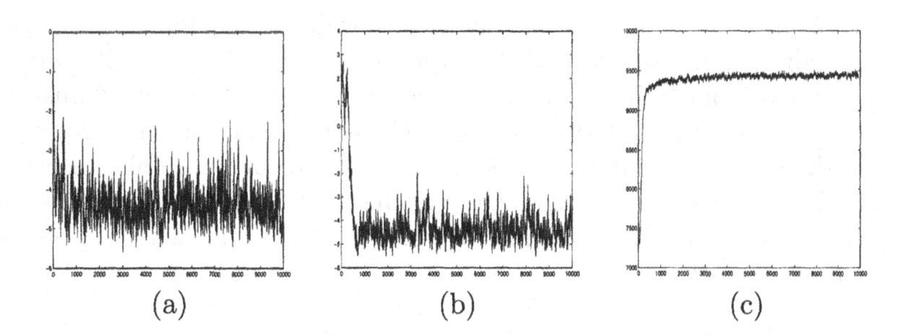

FIGURE 3.20. Convergence diagnostics for the experiment in Figure 3.19. Panels (a) and (b) show trace plots of the log-variance at different positions, while panel (c) shows a trace plot of the functional $f(\boldsymbol{\nu}) = \beta \sum_m \omega_m \sum_i \phi(D_i^{(m)} \boldsymbol{\nu})$.

3.7.3 Contour estimation

In this example we will use a template model for detecting the outline of an artery wall in an ultrasound image. As noted already, this could be used in diagnosing atherosclerosis, since diseased arteries are less likely to dilate in response to infusion of achetylcholine. For this procedure to be useful, we need to quantify the uncertainty of the given answer, for instance by means of an interval estimate. Thus, doing a segmentation of the image, or using a standard contour-detection method, would not be satisfactory, as they only provide point estimates. Moreover, procedures such as segmentation are not robust with respect to image artifacts such as the *missing edge* at the lower right of the artery wall in Figure 3.18b. Such artifacts can easily be dealt with in a template model, see Husby (2001) for details.

We model the artery outline $\boldsymbol{e}$ as the result of applying a transformation to a predefined circular template $\boldsymbol{e}^0$ with m edges. The transformation vector $\boldsymbol{s}$ is modelled as a second order circulant Gaussian Markov random field with precision matrix $\boldsymbol{Q}_s = \boldsymbol{I}_2 \otimes \boldsymbol{Q}$, where $\boldsymbol{Q}$ is a circulant Toeplitz matrix with entries

$$Q_{ij} = \begin{cases} \frac{\kappa}{m} + 6\eta m^3, & j = i \\ -4\eta m^3, & j = i-1, i+1 \bmod m \\ \eta m^3, & j = i-2, i+2 \bmod m, \end{cases} \qquad \kappa, \eta > 0 \tag{3.90}$$

With this parametrisation the behaviour of the model is approximately independent of the number m of edges. We assign independent Gamma priors $\Gamma(a_\kappa, b_\kappa)$ and $\Gamma(a_\eta, b_\eta)$ to the parameters κ and η. See Figure 3.16 for some realisations from this model and Hobolth & Jensen (2000) for some explicit results of the limiting process.

Having an explicit model for the artery wall, we no longer need the implicit edge model, but a model for the log-variance field is still needed. A very simple approach is to assume that there are two smooth Gaussian fields $\boldsymbol{\nu}_0$ and $\boldsymbol{\nu}_1$ associated with the back- and foreground, respectively. The fields are defined on the whole image domain, but are only observed within their respective regions; thus, letting $\mathcal{T}_{\boldsymbol{s}} \subset \mathcal{I}$ be the set of pixels enclosed by the template deformed by $\boldsymbol{s}$, the conditional distribution for the radio frequency field $\boldsymbol{x}$ is

$$x_i \mid \nu_{0,i}, \nu_{1,i}, \boldsymbol{s} \sim \begin{cases} \mathrm{N}\left(0, \exp(-2\nu_{0,i})\right), & i \in \mathcal{T}_{\boldsymbol{s}}^C \cap \mathcal{I} \\ \mathrm{N}\left(0, \exp(-2\nu_{1,i})\right), & i \in \mathcal{T}_{\boldsymbol{s}} \cap \mathcal{I} \end{cases} \qquad \forall i \in \mathcal{I}. \tag{3.91}$$

For simplicity we use an intrinsic Gaussian Markov random field model for the log-variance fields,

$$\pi(\boldsymbol{\nu}_0, \boldsymbol{\nu}_1) \propto \exp\left(-\sum_{k=0}^{1} \tau_k \sum_{i\sim j} q_{ij} \left(\upsilon_{k,i} - \upsilon_{k,j}\right)^2\right), \tag{3.92}$$

where

$$q_{ij} = \begin{cases} |\partial_i|, & i = j \\ -1, & i \sim j, \end{cases}$$

and 0 otherwise. The precisions τ_0 and τ_1 are given Gamma priors with common hyperparameters c and d.

Sampling is again done most simply using single site random walk Metropolis-Hastings algorithms, but experience shows that this leads to slow convergence for our model. Instead we have used a block sampling algorithm, see Husby (2001) for details. Figure 3.21a shows a point estimate of the artery wall based on 200,000 iterations of the sampler. To get a measure of the uncertainty we have plotted samples from the posterior distribution in Figure 3.21b. The variation seems to correspond well with the human perception of uncertainty. A trace plot of the cross-sectional area of the blood vessel is shown in Figure 3.22; the plot indicates that the chain mixes well, and confirms that there is a great deal of uncertainty. A density estimate of the cross-sectional area is shown in Figure 3.23.

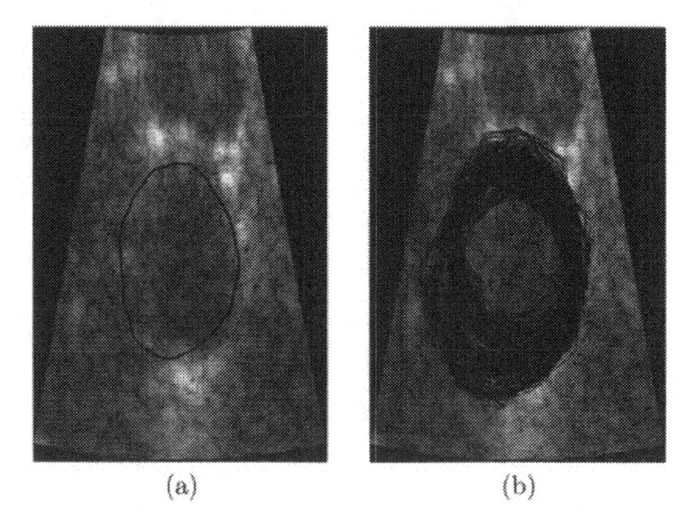

FIGURE 3.21. Contour estimation: (a) Point estimate of the vessel contour; (b) Samples from the posterior distribution, taken with a separation of 500 iterations.

Acknowledgments: Thanks to Adrian Baddeley for providing the newspaper image.

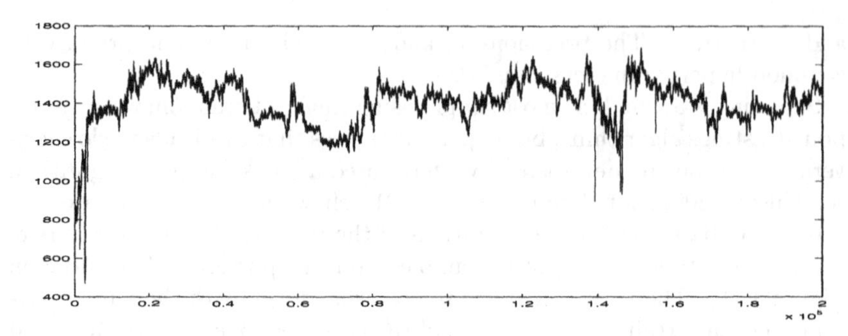

FIGURE 3.22. Trace plot of the cross-sectional area of the template in Figure 3.21.

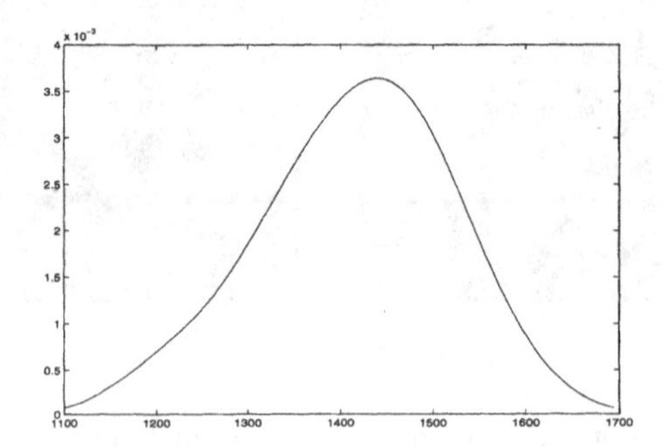

FIGURE 3.23. Density estimate of the cross-sectional area of the blood vessel in Figure 3.18 (a).

3.8 References

Baddeley, A. J. & Van Lieshout, M. N. M. (1993). Stochastic geometry models in high–level vision, *in* K. V. Mardia & G. K. Kanji (eds), *Statistics and Images*, Vol. 20, Carfax Publishing, Abingdon, pp. 235–256.

Besag, J. (1974). Spatial interaction and the statistical analysis of lattice systems (with discussion), *Journal of the Royal Statistical Society, Series B* **36**: 192–225.

Besag, J. (1986). On the statistical analysis of dirty pictures (with discussion), *Journal of the Royal Statistical Society, Series B* **48**: 259–302.

Besag, J. & Green, P. J. (1993). Spatial statistics and Bayesian computation (with discussion), *Journal of the Royal Statistical Society, Series B* **55**: 25–37.

Blake, A. & Isard, M. (1998). *Active Contours*, Springer-Verlag, Berlin.

Blake, A. & Zisserman, A. (1987). *Visual Reconstruction*, MIT Press, Cambridge, MA.

Charbonnier, P. (1994). *Reconstruction d'image: Régularization avec prise en compte des discontinuités*, PhD thesis, Univ. Nice, Sophia Antipolis, France.

Charbonnier, P., Blanc-Feraud, L., Aubert, G. & Barlaud, M. (1997). Deterministic edge-preserving regularization in computed imaging, *IEEE Transaction on Image Processing* **6**: 298–311.

Dryden, I. & Mardia, K. (1999). *Statistical Shape Analysis*, John Wiley and Sons, Chichester.

Friel, N. & Molchanov, I. (1998). Distances between grey-scale images, *Mathematical morphology and its applications to image and signal processing*, Vol. 12 of *Comput. Imaging Vision*, Amsterdam, The Netherlands, pp. 283–290.

Geman, D. (1990). Random fields and inverse problems in imaging, *in* P. L. Hennequin (ed.), *Ecole d'ete de probabilites de Saint-Flour XVIII, 1988*, Springer, Berlin. Lecture Notes in Mathematics, no 1427.

Geman, D. & Yang, C. (1995). Nonlinear image recovery with half-quadratic regularization, *IEEE Transaction on Image Processing* **4**: 932–946.

Geman, S. & Geman, D. (1984). Stochastic relaxation, Gibbs distributions, and the Bayesian restoration of images, *IEEE Transactions on Pattern Analysis and Machine Intelligence* **6**: 721–741.

Geman, S. & McClure, D. (1987). Statistical methods for tomographic image reconstruction, *Proc. 46th Sess. Int. Stat. Inst. Bulletin ISI*, Vol. 52.

Geyer, C. (1999). Likelihood inference for spatial point processes, *in* O. E. Barndorff-Nielsen, W. S. Kendall & M. N. M. van Lieshaut (eds), *Stochastic Geometry: Likelihood and Computation*, Chapman and Hall/CRC, London, Boca Raton, pp. 79–140.

Geyer, C. J. & Møller, J. (1994). Simulation procedures and likelihood inference for spatial point processes, *Scandinavian Journal of Statistics* **21**: 359–373.

Glasbey, C. A. & Mardia, K. V. (2001). A penalized likelihood approach to image warping (with discussion), *Journal of the Royal Statistical Society, Series B* **63**: 465–514.

Green, P. J. (1990). Bayesian reconstruction from emission tomography data using a modified EM algorithm, *IEEE Transaction on Medical Imaging* **9**: 84–93.

Green, P. J. (1995). Reversible jump MCMC computation and Bayesian model determination, *Biometrika* **82**: 711–732.

Greig, D. M., Porteous, B. T. & Scheult, A. H. (1989). Exact maximum a posteriori estimation for binary images, *Journal of the Royal Statistical Society, Series B* **51**: 271–279.

Grenander, U. (1993). *General Pattern Theory*, Oxford University Press, Oxford.

Grenander, U., Chow, Y. & Keenan, D. M. (1991). *Hands: a Pattern Theoretic Study of Biological Shapes*, Research Notes on Neural Computing, Springer, Berlin.

Grenander, U. & Miller, M. I. (1994). Representations of knowledge in complex systems (with discussion), *Journal of the Royal Statistical Society, Series B* **56**: 549–603.

Grimmett, G. (1987). Interacting particle systems and random media: An overview, *International Statistics Review* **55**: 49–62.

Hebert, T. & Leahy, R. (1989). A generalized EM algorithm for 3D Bayesian reconstruction form Poisson data using Gibbs priors, *IEEE Transaction on Medical Imaging* **8**: 194–202.

Hobolth, A. & Jensen, E. B. V. (2000). Modelling stochastic changes in curve shape, with application to cancer diagnostics, *Advances in Applied Probability (SGSA)* **32**: 344–362.

Hokland, J. & Kelly, P. (1996). Markov models of specular and diffuse scattering in restoration of medical ultrasound images, *IEEE Transactions on Ultrasonics Ferroelectrics and Frequency Control* **43**: 660–669.

Husby, O. (2001). High-level models in ultrasound imaging, *Preprint*, Department of Mathematical Sciences, Norwegian University of Technology and Science, Trondheim.

Husby, O., Lie, T., Langø, T., Hokland, J. & Rue, H. (2001). Bayesian 2d deconvolution: A model for diffuse ultrasound scattering, *IEEE Transactions on Ultrasonics Ferroelectrics and Frequency Control* **48**: 121–130.

O'Rourke, J. (1998). *Computational Geometry in C*, Cambridge University Press, Cambridge.

Rue, H. (2001). Fast sampling of Gaussian Markov random fields with applications, *Journal of the Royal Statistical Society, Series B* **63**: 325–338.

Rue, H. & Hurn, M. A. (1999). Bayesian object identification, *Biometrika* **86**: 649–660.

Stander, J. & Silverman, B. W. (1994). Temperature schedules for simulated annealing, *Statistics and Computing* **4**: 21–32.

Swendsen, R. & Wang, J. (1987). Nonuniversal critical dynamics in Monte Carlo simulations, *Physical Review Letters* **58**: 86–88.

Tjelmeland, H. & Besag, J. (1998). Markov random fields with higher order interactions, *Scandinavian Journal of Statistics* **25**: 415–433.

Van Lieshout, M. N. M. (1995). Markov point processes and their applications in high-level imaging (with discussion), *Bulletin of the International Statistical Institute* **LVI, Book 2**: 559–576.

Winkler, G. (1995). *Image Analysis, Random Fields and Dynamic Monte Carlo Methods*, Springer, Berlin.

Wood, A. T. A. & Chan, G. (1994). Simulation of stationary Gaussian processes in $[0,1]^d$, *Journal of Computational and Graphical Statistics* **3**: 409–432.

4

An Introduction to Simulation-Based Inference for Spatial Point Processes

Jesper Møller
Rasmus P. Waagepetersen

4.1 Introduction

Spatial point processes play a fundamental role in spatial statistics. In the simplest case they model "small" objects that may be identified by a map of points showing stores, towns, plants, nests, or cases of a disease observed in a two dimensional region or galaxies observed in a three dimensional region. The points may be decorated with marks (such as sizes or types) whereby marked point processes are obtained. The areas of applications are manifold: astronomy, geography, ecology, forestry, spatial epidemiology, image analysis, and many more. Currently spatial point processes is an active area of research, which probably will be of increasing importance for many new applications.

This chapter aims at collecting some of the recent theoretical advances and examples of applications in simulation-based inference for spatial point processes in a concise manner. We focus mainly on general ideas and methodology without going too much into technical details. Just about all the material, and more, will appear in Møller & Waagepetersen (2003). Sometimes the exposition will be biased towards our own work and interests, but most of the time we follow the mainstream of recent contributions in this research area. Particularly we focus on Cox processes and Markov point processes, which we find are the most useful models in data analysis. Moreover, we relax the stationary assumption, which seems very timely.

The chapter is organised as follows. Section 4.2 introduces two examples of applications which are used throughout this chapter. Section 4.3 describes in more detail what is meant by a spatial point process. Section 4.4 deals with Poisson point processes. Section 4.5 surveys explanatory analysis and model validation based on various kind of summary statistics. Sections 4.6–4.7 are the two main sections. Section 4.6 considers cluster processes and Cox processes, particularly log Gaussian Cox processes, and discusses simulation-based inference for clustered point patterns. Section 4.7 deals with different aspects of model construction, simulation (including perfect simulation), and inference for Markov point processes. Finally, Section 4.8 contains some concluding remarks.

4.2 Illustrating examples

Examples 1 and 2 below are used for illustrative purposes throughout this chapter. In Example 1 we consider what is later on called a simple point process, while the discs in Example 2 will later on be treated as a so-called marked point process.

4.2.1 Example 1: Weed plants

Figure 4.1 shows the position of 976 weed plants (*Trifolium* spp./clover) observed within 45 metal frames on a Danish barley field. This point pattern is a subset of a much larger dataset analysed in Brix & Møller (2001) and Brix & Chadoeuf (2000) where several weed species at different dates were considered. Note that we have rotated the design 90° in Figure 4.1. The 45 frames are of size 30×20 cm^2, and they are organised in 9 groups each containing 5 frames, where the vertical and horizontal distances between two neighbouring groups are 1 m and 1.5 m, respectively. The size of the experimental area is 7.5×5 m^2, where the longest side agrees with the ploughing direction. Note the trend in the point pattern: in general more weed plants occur in the upper frames (i.e. the frames to the left in Figure 4.1).

4.2.2 Example 2: Norwegian spruces

The Norwegian spruce data (Fiksel 1984, Penttinen, Stoyan & Henttonen 1992, Stoyan, Kendall & Mecke 1995, Goulard, Särkkä & Grabarnik 1996) is an example of a pattern of discs. The data are shown in Figure 4.2, where the centres of the 134 discs are the positions of the spruces observed in a rectangular window of size 56×38 m^2, and the radii are the stem diameters multiplied by 5. As discussed in Penttinen et al. (1992) and Goulard et al. (1996) the "influence zone" of a tree is about 5 times the stem diameter.

4.3 What is a spatial point process?

A formal answer to this question is given below by considering in Section 4.3.1 the special case of simple point processes in $\mathbb{R}^d$, in Section 4.3.2

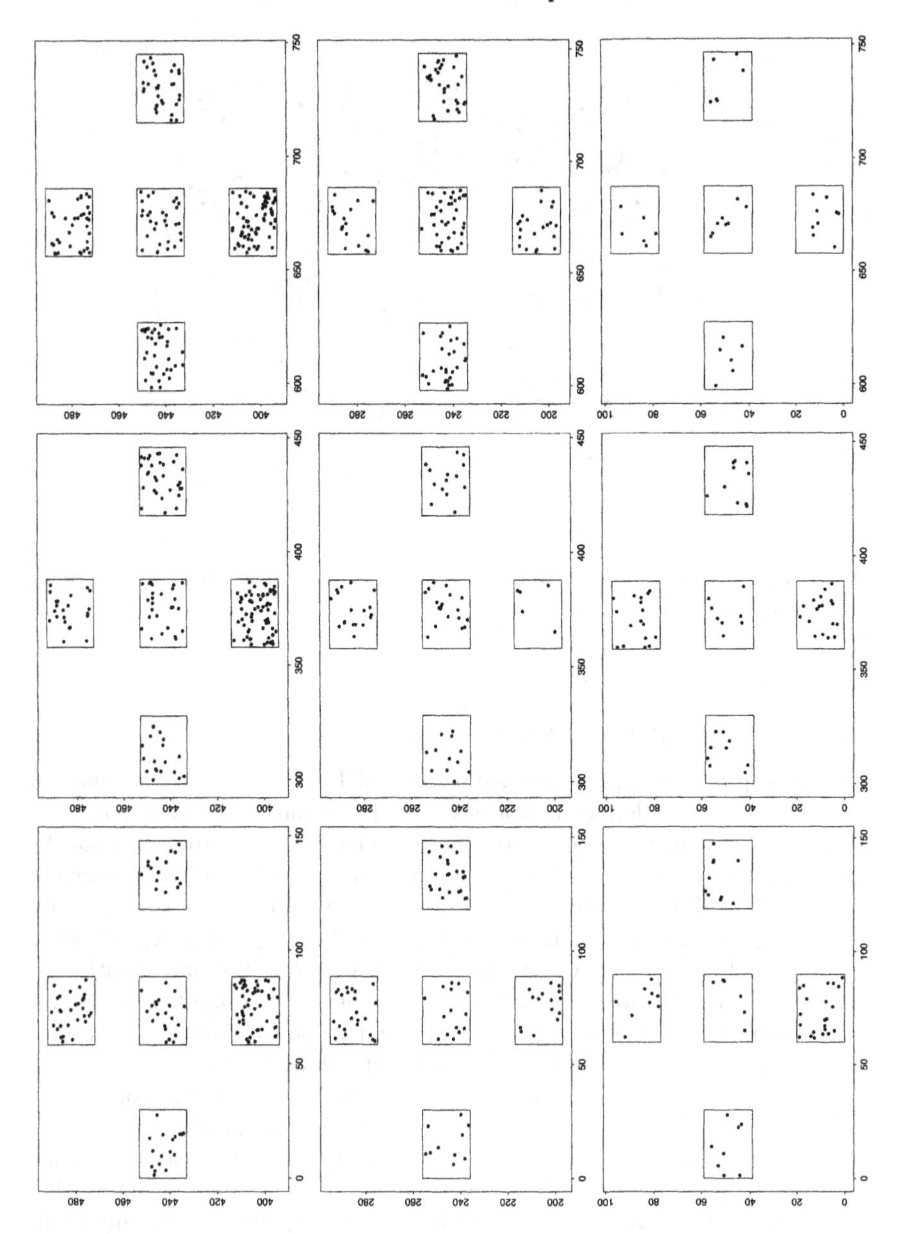

FIGURE 4.1. Positions of weed plants when the design is rotated 90°.

some special cases of marked point processes which all can be related to the examples in Section 4.2, and in Section 4.3.3 a general setting and notation used throughout this text.

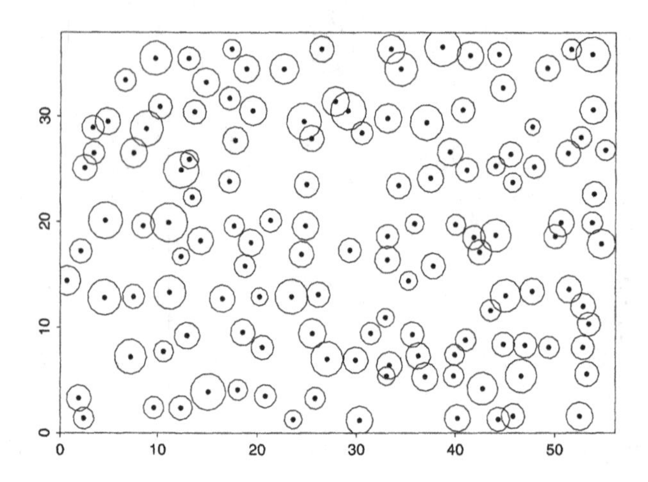

FIGURE 4.2. Positions of Norwegian spruces. The radii of the discs equal 5 times the stem diameters.

4.3.1 Simple point processes in $\mathbb{R}^d$

A simple point process in the d-dimensional Eucledian space $\mathbb{R}^d$ may be considered as a random countable set $X \subset \mathbb{R}^d$. What is meant here by random will be made precise in a moment. The elements in X are called *points*, and they may represent the occurrence of some event like the occurrence of a weed plant. Equivalently we may view X as a counting measure defined by $X(A) = n(X_A)$ for bounded Borel sets $A \subset \mathbb{R}^d$, where $n(X_A)$ denotes the cardinality of $X_A = X \cap A$. We here for simplicity restrict attention to the case of *simple point processes*, i.e. when there are no multiple points; a counting measure with multiplicities may in fact be considered as a special case of a marked point process as described below. Moreover, we exclude the case of accumulating points, i.e. $X(A) < \infty$ whenever A is bounded.

In practice $X = X_S$ will be concentrated on a bounded set $S \subset \mathbb{R}^d$, but for mathematical convenience, or if S is large or unknown, we may let $S = \mathbb{R}^d$. Usually in applications S is d-dimensional (we may formalize this by saying that S is topologically regular, i.e. it is equal to the closure of its interior), and most examples in the point process literature concerns the planar case $d = 2$. In other cases S may be a lower-dimensional manifold, e.g. the $(d-1)$-dimensional unit sphere. In practice we observe only X_W where $W \subseteq S$ is some bounded observation window like in Figures 4.1 and 4.2. If X is not concentrated on W, it may be important to take *boundary or edge effects* into account in the statistical analysis, as the unobserved points outside W may affect X_W.

4.3.2 Marked point processes in $\mathbb{R}^d$

Suppose now that Y is a simple point process in $\mathbb{R}^d$ and a random "mark" $m_\xi \in \mathcal{M}$ is attached to each point $\xi \in Y$. Then $X = \{(\xi, m_\xi) : \xi \in Y\}$ is called a *marked point process* in $\mathbb{R}^d$ with *mark space* $\mathcal{M}$. The marks may be dependent or not of each other and of Y; again what is meant by randomness is made precise below.

One simple example is a disc process as considered in Example 2 (Norwegian spruces), letting $\mathcal{M} = (0, \infty)$ and identifying (ξ, m_ξ) with the disc with centre ξ and radius m_ξ. Similarly, we obtain marked point processes for other kinds of geometric objects (line segments, ellipses, etc.), also called *germ-grain models* where ξ (the germ) specifies the location of the object m_ξ (the grain). See Section 3.6.3.

Another example is a *multivariate* or *multi-type point process*, where $\mathcal{M} = \{1, \ldots, k\}$ and the marks specify k different types of points (e.g. different types of weed plants).

4.3.3 General setting and notation

Many readers (particularly those who are unfamiliar with measure theory) may prefer to skip the general setting described below. However, we recommend the reader to at least notice the meaning of the following notation which is used throughout this chapter.

Notation

We consider X to be a locally finite subset of a rather general metric space S with metric $d(\cdot,\cdot)$. Here locally finiteness means that $X(A) = n(X_A) < \infty$ for any bounded set $A \subseteq S$. The state space of X is denoted N and consists of all locally finite point configurations in S:

$$N = \{x \subset S : x(A) < \infty \text{ for all } A \in \mathcal{B}_0\}$$

where $\mathcal{B}_0$ denotes the class of bounded Borel sets contained in S. This state space is equipped with a suitable σ-algebra denoted $\mathcal{N}$ (see below). Elements of S and N are usually denoted $\xi, \eta, \ldots \in S$ and $x, y, \ldots \in N$. We abuse the notation and write $x \cup \xi$ for $x \cup \{\xi\}$ (with $x \in N$ and $\xi \in S \setminus x$), $y \setminus \eta$ for $y \setminus \{\eta\}$ (with $y \in N$ and $\eta \in S$), etc.

Measure theoretical details

Formally, we assume S to be a Polish space, i.e. a complete separable metric space; though this assumption may be weakened, it is commonly satisfied in applications and ensures the validity of some desirable properties as listed below. Moreover, we assume X to be defined on some underlying probability space $(\Omega, \mathcal{F}, P)$ so that measurability of X means that $X(A) : \Omega \to \mathbb{N}$ is a measurable function whenever $A \in \mathcal{B}_0$. In other words, we equip N with the smallest σ-algebra $\mathcal{N}$ containing all sets of the form $\{x \in N : x(A) = m\}$ with $A \in \mathcal{B}_0$ and $m \in \mathbb{N}_0$. It can be shown (Matheron 1975, Daley & Vere-Jones 1988) that

- N is a metric space, $\mathcal{N}$ is the corresponding Borel σ-algebra, and $\mathcal{N}$ is countably generated;
- the distribution of X is determined by the *void probabilities* as given by $P(X(A) = 0)$, $A \in \mathcal{B}_0$.

How is this related to simple and marked point processes considered so far?

For a simple point process in $\mathbb{R}^d$ we let $d(\cdot,\cdot)$ be the usual Eucledian metric or distance $\|\cdot\|$.

A marked point process with locations in $\mathbb{R}^d$ and mark space $\mathcal{M}$ can be considered as a point process defined on $S = \mathbb{R}^d \times \mathcal{M}$. Then S becomes a Polish space, if we assume the mark space $\mathcal{M}$ to be a Polish space with metric $d_{\mathcal{M}}$, and equip S with the metric

$$d((\xi_1, m_{\xi_1}), (\xi_2, m_{\xi_2})) = \max\{\|\xi_1 - \xi_2\|, d_{\mathcal{M}}(m_{\xi_1}, m_{\xi_1})\}.$$

This is also a natural metric in the sense that the Borel σ-algebra for S agrees with the product σ-algebra of the Borel sets in $\mathbb{R}^d$ and $\mathcal{M}$. If the marks are geometrical objects, we may use the formalism of stochastic geometry, considering $\mathcal{M}$ as the space of compact subsets of $\mathbb{R}^d$ equipped with the Hausdorff metric defined by

$$d_{\mathcal{M}}(A, B) = \inf\{r \geq 0 : A \subset B \oplus b(0,r),\ B \subset A \oplus b(0,r)\}$$

where $\oplus$ denotes Minkowski-addition, i.e. $A \oplus B = \{a + b : a \in A, b \in B\}$. Then M is a Polish space (Matheron 1975). For example for balls $A = b(0, r_1)$ and $B = b(0, r_2)$ with centre 0 and radii $0 < r_1 \leq r_2 < \infty$, $d_{\mathcal{M}}(A, B) = r_2 - r_1$.

4.4 Poisson point processes

Poisson point processes play a fundamental role, as they serve as a tractable model class for "no interaction" or "complete spatial randomness" in spatial point patterns, and as reference processes when comparing and constructing more advanced point process models. Furthermore, the initial step of a point process analysis often consists of looking for discrepancies with a Poisson model. This may point to alternative models as discussed later in Section 4.5.

General definitions and properties of Poisson processes are reviewed in Section 4.4.1, the particular case of Poisson processes in $\mathbb{R}^d$ is considered in Section 4.4.2, and Section 4.4.3 concerns marked Poisson processes in $\mathbb{R}^d$. Further material can be found in Daley & Vere-Jones (1988), Kingman (1993), and Stoyan et al. (1995).

4.4.1 Definitions and properties

Let μ be a locally finite and diffuse measure defined on the Borel sets in S, i.e. $\mu(A) < \infty$ for $A \in \mathcal{B}_0$ and μ has no mass at any point in S. We say that X is a *Poisson point process* on S with *intensity measure* μ, and write

$$X \sim \text{Poisson}(S, \mu),$$

if X satisfies the following properties:

- for any $A \in \mathcal{B}_0$, $X(A) \sim \text{po}(\mu(A))$, the Poisson distribution with mean $\mu(A)$;
- *independent scattering:* for any disjoint Borel sets $A_1, \ldots, A_n \subseteq S$ with an arbitrary $n \geq 2$, $X(A_1), \ldots, X(A_n)$ are independent.

The assumption that μ is locally finite and diffuse ensures that X is a locally finite point process with no multiple points. The independent scattering property explains the terminology of "no interaction" and "complete spatial randomness". In the definition above we can replace the independent scattering property by

- for any $A \in \mathcal{B}_0$ with $\mu(A) > 0$, conditional on $X(A) = n$, the n points in X_A are mutually independent with common distribution $\bar{\mu}(\cdot) = \mu(\cdot \cap A)/\mu(A)$; this is called a *binomial point process* of n points with distribution $\bar{\mu}$.

It is therefore not hard to verify that there exists a well-defined point process with these properties.

The simplest way of characterising a Poisson point process is by its void probabilities,

$$P(X(A) = 0) = \exp(-\mu(A)), \quad A \in \mathcal{B}_0.$$

A less well-known but very useful characterisation of a Poisson process is provided by the *Slivnyak-Mecke theorem* (Mecke 1967): $X \sim \text{Poisson}(S, \mu)$ if and only if, for any measurable function $h : N \times S \to [0, \infty)$,

$$\mathbb{E} \sum_{\xi \in X} h(X \setminus \xi, \xi) = \int \mathbb{E} h(X, \xi) \mu(\mathrm{d}\xi). \tag{4.1}$$

The extended Slivnyak-Mecke theorem is obtained by induction: we have that $X \sim \text{Poisson}(S, \mu)$ if and only if, for any $n \in \mathbb{N}$ and any measurable function $h : N \times S^n \to [0, \infty)$,

$$\mathbb{E} \sum_{\xi_1,\ldots,\xi_n \in X}^{\neq} h(X \setminus \{\xi_1, \ldots, \xi_n\}, \xi_1, \ldots, \xi_n) = \int \cdots \int \mathbb{E} h(X, \xi_1, \ldots, \xi_n) \mu(\mathrm{d}\xi_1) \cdots \mu(\mathrm{d}\xi_n), \tag{4.2}$$

where the $\neq$ over the summation sign means that the n points $\xi_1, \ldots, \xi_n$ are all different.

The class of Poisson processes is closed under two basic operations for point processes:

- *superpositioning:* if $X_1 \sim \text{Poisson}(S, \mu_1)$ and $X_2 \sim \text{Poisson}(S, \mu_2)$ are independent, then $X_1 \cup X_2 \sim \text{Poisson}(S, \mu_1 + \mu_2)$;
- *independent thinning:* if $X \sim \text{Poisson}(S, \mu)$ and $R(\xi) \sim \text{Uniform}[0, 1]$, $\xi \in S$, are mutually independent, and $p : S \to [0, 1]$ is a measurable function, then $Z = \{\xi \in X : R(\xi) < p(\xi)\} \sim \text{Poisson}(S, \nu)$ with $\nu(A) = \int_A p(\xi)\mu(\mathrm{d}\xi)$.

These statements are easily verified by considering the void probabilities of the superposition $X_1 \cup X_2$ and the thinned process Z.

4.4.2 Poisson processes in $\mathbb{R}^d$

Suppose that $X \sim \text{Poisson}(\mathbb{R}^d, \mu)$. If μ is absolutely continuous with respect to the Lebesgue measure, then its density $\rho(\xi) = \mathrm{d}\mu(\xi)/\mathrm{d}\xi$ is called the *intensity function.* Often in statistical modelling of a Poisson point process, one specifies a parametric model for the intensity function, cf. Section 4.5.5. This may depend on covariate information as e.g. in Rathbun (1996).

In the particular case where $\rho(\cdot) = \rho$ is constant, X is said to be a *homogeneous Poisson point process with intensity* ρ. This is equivalent to assuming *stationarity* of X under translations, that is the distribution of $X + s = \{\xi + s : \xi \in X\}$ is the same as that of X for any $s \in \mathbb{R}^d$. A homogeneous Poisson point process is also *isotropic* as its distribution is invariant under rotations in $\mathbb{R}^d$. In the special case $d = 1$, the distances between successive points of a homogeneous Poisson point process are independent and exponentially distributed with mean $1/\rho$.

Consider a Poisson point process X with intensity function ρ. It is in general impossible to specify the density of X with respect to another Poisson process. For example, a homogeneous Poisson point process with intensity $\rho_1 > 0$ is absolutely continuous with respect to another homogeneous Poisson point process with intensity $\rho_2 > 0$ if and only if $\rho_1 = \rho_2$. If $W \subset \mathbb{R}^d$ has Lebesgue measure (or area/volume) $|W| \in (0, \infty)$ and $\int_W \rho(\xi)\mathrm{d}\xi < \infty$, then X_W has a *density*

$$f_W(x) = \exp\left(|W| - \int_W \rho(\xi)\mathrm{d}\xi\right) \prod_{\xi \in x} \rho(\xi), \quad x \subset W,\ x(W) < \infty, \tag{4.3}$$

with respect to the *standard Poisson point process* Poisson($\mathbb{R}^d$, Lebesgue).

Simulation of a homogeneous Poisson point process X with intensity $\rho > 0$ within a d-dimensional box $B = [0, a_1] \times \cdots \times [0, a_d]$ is straightforward: first generate $N \sim \mathrm{po}(\rho a_1 \cdots a_d)$ and next N independent and uniformly distributed points in B. Alternatively, we may use that

- the first coordinates $\xi^{(1)}$ of points $\xi = (\xi^{(1)}, \xi^{(2)}) \in X$ with $\xi^{(2)} \in [0, a_2] \times \cdots \times [0, a_d]$ form a homogeneous Poisson process on the real line with intensity $\rho a_2 \cdots a_d$,
- the remaining components $\xi^{(2)}$ of such points are independent and uniformly distributed on $[0, a_2] \times \cdots \times [0, a_d]$.

For simulation within a ball in $\mathbb{R}^d$, it is more convenient to make a shift to polar coordinates and use a radial simulation procedure (Quine & Watson 1984).

Combining this with independent thinning we obtain a simple simulation procedure for inhomogeneous Poisson processes with an intensity function $\rho(\cdot)$ which is bounded by a constant c on $B \subset \mathbb{R}^d$: generate a homogeneous Poisson process on B with intensity c, and let the retention probabilities be $p(\xi) = \rho(\xi)/c$, $\xi \in B$.

4.4.3 Marked Poisson processes in $\mathbb{R}^d$

Suppose that $X \sim \mathrm{Poisson}(\mathbb{R}^d \times \mathcal{M}, \mu)$. Independence between points and marks in X is equivalent to that μ factorises into a product measure $\mu = \nu \times Q$ where ν is a locally finite measure in $\mathbb{R}^d$ and Q is a probability measure describing the common distribution of the marks; it is simply called the *mark distribution*. We shall not pay much attention to marked Poisson processes, but refer the interested reader to Stoyan et al. (1995) and the references therein.

4.5 Summary statistics

Exploratory analysis for spatial point patterns and the validation of fitted models are often based on non-parametric estimates of various summary statistics, cf. e.g. Ripley (1977), Stoyan et al. (1995), and Ohser & Mücklich (2000). In this section we confine ourselves to summary statistics for a single point pattern X observed in a bounded planar window $W \subset \mathbb{R}^2$

with Lebesgue measure $|W| > 0$. Extensions to replicated point patterns and to marked point processes are sometimes obvious; see Diggle, Lange & Beneš (1991), Diggle, Mateu & Clough (2000), Baddeley, Moyeed, Howard & Boyde (1993), Schlather (2001), and the references therein.

Sections 4.5.1–4.5.2 consider summary statistics related to the first and second order moments of the counts $X(A)$, $A \in \mathcal{B}_0$, while summary statistics based on distribution functions for interpoint distances are treated in Section 4.5.3.

4.5.1 First order characteristics

Just as for Poisson point processes in $\mathbb{R}^d$, we define the following concepts. The *intensity measure* μ of X is given by $\mu(A) = \mathbb{E}X(A)$ for Borel sets $A \subseteq \mathbb{R}^d$. If μ is absolutely continuous with respect to the Lebesgue measure, its density $\rho(\xi) = \mathrm{d}\mu(\xi)/\mathrm{d}\xi$ is called the *intensity function*. Loosely speaking, $\rho(\xi)\mathrm{d}\xi$ is the probability for the occurrence of a point in an infinitesimally small ball with centre ξ and area $\mathrm{d}\xi$. If moreover $\rho(\xi) = \rho$ is constant, X is said to be *homogeneous or first order stationary with intensity* ρ; otherwise X is said to be *inhomogeneous*. Clearly, stationarity of X under translations implies homogeneity of X.

In the homogeneous case, a natural unbiased estimator is $\hat{\rho} = X(W)/|W|$. This is in fact the maximum likelihood estimator if X is a homogeneous Poisson process.

In the inhomogeneous case, a non-parametric kernel estimator of the intensity function (assuming this exists) is

$$\hat{\rho}(\xi) = \sum_{\eta \in X_W} k_1(\xi - \eta)/c_{W,k_1}(\eta), \quad \eta \in W, \tag{4.4}$$

where k_1 is a kernel (density function) and $c_{W,k_1}(\eta) = \int_W k_1(\xi - \eta)\mathrm{d}\xi$ is an *edge correction factor* so that $\int_W \hat{\rho}(\xi)\mathrm{d}\xi$ is an unbiased estimator of $\mu(W)$ (Diggle 1985). The estimator is usually very sensitive to the choice of bandwidth, while the choice of kernel function is less important.

Alternative estimators in connection to improved ratio-type estimators are discussed in Section 4.5.2, cf. (4.10) and (4.12).

4.5.2 Second order characteristics

Pair correlation, K, and L-functions

The so-called *second order factorial moment measure* is given by

$$\alpha^{(2)}(A \times B) = \mathbb{E} \sum_{\xi, \eta \in X}^{\neq} 1[\xi \in A, \eta \in B] = \mathbb{E}\big[X(A)X(B)\big] - \mu(A \cap B) \tag{4.5}$$

for Borel sets $A, B \subseteq \mathbb{R}^d$, where $1[\cdot]$ denotes indicator function. When X is Poisson with intensity measure μ, combining (4.2) and (4.5), we obtain that $\alpha^{(2)}(A \times B) = \mu(A)\mu(B)$. If $\alpha^{(2)}$ has a density $\rho^{(2)}(\xi, \eta)$ with respect to the Lebesgue measure on $\mathbb{R}^d \times \mathbb{R}^d$, this is called the *second order product density*; intuitively, $\rho^{(2)}(\xi, \eta)\mathrm{d}\xi\mathrm{d}\eta$ is the probability for observing a point in each of the infinitesimally small balls with centres ξ, η and areas $\mathrm{d}\xi, \mathrm{d}\eta$, respectively.

A widely used summary statistic (in spatial statistics and particularly astronomy and astrophysics, see e.g. Peebles (1974) and Kerscher (2000)), is the *pair correlation function* given by

$$g(\xi, \eta) = \rho^{(2)}(\xi, \eta)/(\rho(\xi)\rho(\eta))$$

provided the terms on the right hand side exist. For a Poisson process, we have that $g = 1$. In general, $g(\xi, \eta) > 1$ indicates *attraction* or *clustering*, and $g(\xi, \eta) < 1$ *repulsion* or *regularity* for points at locations ξ, η; this may in turn be due to certain latent processes (Section 4.6) or interaction between the points (Section 4.7). It is often assumed that $g(\xi, \eta) = g(\xi - \eta)$ is translation invariant; this is e.g. implied by stationarity of X under translations. It is convenient if $g(\xi, \eta) = g(\|\xi - \eta\|)$ depends only on the distance $\|\xi - \eta\|$; this is the case if X is both stationary and isotropic.

In the stationary case of X with finite intensity $\rho > 0$, there is a close relationship between g and *Ripley's K-function* (Ripley 1977) defined by

$$K(r) = \mathbb{E} \sum_{\xi \in X_W, \eta \in X}^{\neq} 1[\|\xi - \eta\| \leq r]/(\rho^2 |W|). \tag{4.6}$$

It is easily seen that this definition does not depend on W. Further $\rho K(r)$ has an interpretation as the mean number of further points within distance r from a "typical" point in X. If $g(\xi, \eta) = g(\xi - \eta)$ exists and is translation invariant, then

$$K(r) = \int_{\|\xi\| \leq r} g(\xi)\mathrm{d}\xi, \quad r \geq 0.$$

Especially, for a homogeneous Poisson process,

$$K(r) = \omega_d r^d, \quad r \geq 0$$

where $\omega_d = \pi^{d/2}/\Gamma(1 + d/2)$ is the volume of a unit ball. One often considers the *L-function* given by $L = (K/\omega_d)^{1/d}$ instead of K, as $L(r) = r$ is the identity if X is a homogeneous Poisson process, and since this transformation is variance stabilising when K is estimated by non-parametric methods (Besag 1977b).

The K-function can be modified to directional K-functions for the anisotropic case (Stoyan & Stoyan 1994). It can also be extended to the inhomogeneous case (Baddeley, Møller & Waagepetersen 2000), where $K_{\mathrm{inhom}}(r)$

is defined as in (4.6) with ρ^2 replaced by $\rho(\xi)\rho(\eta)$, provided that $K_{\text{inhom}}(r)$ does not depend on W. For example, this assumption is satisfied if $g(\xi, \eta) = g(\xi - \eta)$ exists and is translation invariant, in which case $K_{\text{inhom}}(r) = \int_{\|\xi\| \leq r} g(\xi) \mathrm{d}\xi$. It is also satisfied if X is obtained by independent thinning of a stationary point process. As in the homogeneous case, we often use $L_{\text{inhom}} = (K_{\text{inhom}}/\omega_d)^{1/d}$ instead of K_{inhom}. Note that $L_{\text{inhom}}(r) = r$ in the Poisson case.

The summary statistics g and K_{inhom} are invariant under independent thinning. This also holds for K when all thinning probabilities $p(\xi)$ are equal. Furthermore, after independent thinning of a stationary point process, K_{inhom} for the thinned process agrees with K for the original process. These invariance properties can be exploited for semi-parametric inference, cf. Baddeley et al. (2000).

The summary statistics considered so far describe the second order properties of a spatial point process. It should be noticed that very different point process models can share the same first and second order properties as discussed in Baddeley & Silverman (1984) and Baddeley et al. (2000).

Non-parametric estimation

Non-parametric estimation of summary statistics is discussed in Stoyan & Stoyan (1994), Stoyan & Stoyan (2000), Ohser & Mücklich (2000), and the references therein. Such estimators may take boundary effects into consideration as demonstrated below.

One commonly used estimator of $\rho^2 K(r)$ is

$$\widehat{\rho^2 K(r)} = 2 \sum_{\{\xi,\eta\} \subseteq X_W}^{\neq} 1[\|\xi - \eta\| \leq r] / |W_\xi \cap W_\eta| \tag{4.7}$$

where $W_\xi = \{\xi + \eta : \eta \in W\}$ denotes the translate of W by ξ. Because of stationarity, the estimator is unbiased when

$$|W \cap W_\xi| > 0 \text{ for all } \xi \in \mathbb{R}^d \text{ with } \|\xi\| \leq r. \tag{4.8}$$

For example, if W is rectangular, it is by (4.8) required that r is smaller than the smallest side in W. If the pair correlation of X exists, then the estimator is still unbiased if

$$|W \cap W_\xi| > 0 \text{ for Lebesgue almost all } \xi \in \mathbb{R}^d \text{ with } \|\xi\| \leq r. \tag{4.9}$$

In Section 4.5.5 we compare the conditions (4.8) and (4.9) for the special design in the weed plant example.

The estimator (4.7) may be combined with an estimator of ρ^2 to obtain an estimator of $K(r)$, but the best choice of combination depends on the particular model of X (Stoyan & Stoyan 2000). For example, for a homogeneous Poisson process X, it is recommended to use

$$\widehat{\rho^2} = X(W)(X(W) - 1)/|W|^2 \tag{4.10}$$

which is an unbiased estimator of ρ^2. Then the combined estimator of K is biased but it is said to be *ratio-unbiased.*

A similar situation is noticed in Baddeley et al. (2000) for the inhomogeneous case, where $K_{\text{inhom}}(r)$ is estimated by

$$\widehat{K}_{\text{inhom}}(r) = 2 \sum_{\{\xi,\eta\}\subseteq X_W}^{\neq} 1[\|\xi - \eta\| \leq r]/\{|W_\xi \cap W_\eta|\bar{\rho}(\xi)\bar{\rho}(\eta)\}, \tag{4.11}$$

and instead of the estimator $\hat{\rho}$ in (4.4),

$$\bar{\rho}(\xi) = \sum_{\eta \in X_W \setminus \xi} k_1(\xi - \eta)/c_{W,k_1}(\eta), \quad \eta \in W, \tag{4.12}$$

is used. From (4.11) we obtain an estimator of L_{inhom}.

Non-parametric kernel estimators of g may be derived along similar lines when $g(\xi, \eta) = g(\|\xi - \eta\|)$ depends only on the distance. In the planar case $d = 2$, if $\widehat{\rho(\xi)\rho(\eta)}$ is an estimate of $\rho(\xi)\rho(\eta)$, we may estimate $g(r)$, $r > 0$, in analogy with (4.7) and (4.11) by

$$\hat{g}(r) = 2 \sum_{\{\xi,\eta\}\subseteq X_W}^{\neq} k_2(r - \|\xi - \eta\|) \Big/ \left[\pi r |W_\xi \cap W_\eta| \widehat{\rho(\xi)\rho(\eta)}\right] \tag{4.13}$$

where k_2 is a symmetric kernel. Like in (4.4) the choice of bandwidth is important, and the estimator may be unreliable at small distances r as discussed later in Example 4.5.4. Alternative estimators are discussed in Stoyan & Stoyan (1994), Stoyan & Stoyan (2000), and Ohser & Mücklich (2000).

Plots of these estimators are often supplied with *envelopes* obtained by simulation of a specified model, for example, an estimated Poisson model; several examples are shown in the sequel. Let $\hat{T}_0$ be a non-parametric estimator of a summary statistic T computed from the data X, and let $\hat{T}_1, \ldots, \hat{T}_n$ be the same non-parametric estimator computed from i.i.d. simulations $X_1, \ldots, X_n$ under the specified model. For example, we could have $\hat{T}_0 = \hat{K}(r)$ for a given distance $r > 0$. We have that with probability (at least) $(n-1)/(n+1)$,

$$\min_{1\leq i\leq n} \hat{T}_i \leq \hat{T}_0 \leq \max_{1\leq i\leq n} \hat{T}_i \tag{4.14}$$

if X follows the specified model (strictly speaking this has to be a simple hypothesis — the probability may differ from $(n-1)/(n+1)$ if the "specified model" is actually an estimated model — but this is usually ignored in applications). We refer to the bounds in (4.14) as lower and upper envelopes. In our examples we choose $n = 39$ so that (4.14) specifies a 2.5% lower envelope and a 97.5% upper envelope (for each fixed value of r if e.g. $\hat{T}_0 = \hat{K}(r)$).

Higher order intensities

Finally, we remark that higher order summary statistics can be introduced as well, but the corresponding non-parametric estimators may be less stable if the number of points observed is not sufficiently large; see Peebles & Groth (1975), Stoyan & Stoyan (1994), Møller, Syversveen & Waagepetersen (1998), and Schladitz & Baddeley (2000).

4.5.3 Nearest-neighbour and empty space functions

Consider again the stationary case with finite intensity $\rho > 0$. The *empty space function* (or spherical contact distribution function) F is the distribution function of the distance from the origin (or another fixed point in $\mathbb{R}^d$) to the nearest point in X, i.e.

$$F(r) = P(\inf_{\xi \in X} \|\xi\| \leq r).$$

The *nearest-neighbour function* is defined by

$$G(r) = \mathbb{E} \sum_{\xi \in X_W} 1[\inf_{\eta \in X \setminus \xi} \|\xi - \eta\| \leq r]/(\rho|W|)$$

and has the interpretation as the distribution function of the distance from a typical point in X to its nearest point in X. This definition does not depend on the choice of W. It is not obvious how to extend the definitions of F and G to the inhomogeneous case. For a homogeneous Poisson process, by (4.1), $F(r) = G(r) = 1 - \exp(-\rho\omega_d r^d)$. For other kind of models, closed form expressions of F and G are rarely known.

Van Lieshout and Baddeley (1996) suggest to consider the combined summary statistic

$$J(r) = (1 - G(r))/(1 - F(r)) \quad \text{for } F(r) < 1,$$

which is 1 for a homogeneous Poisson process, whilst values less (more) than 1 may be an indication of clustering (regularity) in X.

Non-parametric estimators of F and G (and thereby J) are easily derived using *minus sampling*: For each $r > 0$, let $W_{\ominus r} = \{\xi \in W : b(\xi, r) \subseteq W\}$ denote the set of points in W with a distance to the boundary of W which is greater than r. If $I_r \subset W_{\ominus r}$ is a finite grid of n_r points (chosen independently of X), we have the unbiased estimator

$$\hat{F}(r) = \sum_{\xi \in I_r} 1[\inf_{\eta \in X} \|\xi - \eta\| \leq r]/n_r \tag{4.15}$$

and the ratio-unbiased estimator

$$\hat{G}(r) = \sum_{\xi \in X_{W_{\ominus r}}} 1[\inf_{\eta \in X \setminus \xi} \|\xi - \eta\| \leq r]/(\hat{\rho}|W_{\ominus r}|). \tag{4.16}$$

Finally, envelopes may be simulated in the same way as described above for the L-function.

Alternative Kaplan-Meyer type estimators are discussed in Baddeley & Gill (1997).

4.5.4 Example 2: Norwegian spruces (continued)

In this section we consider only the positions of the spruces.

A non-parametric kernel estimate (4.4) of the intensity surface for the spruce positions is shown in Figure 4.3. A Gaussian kernel k_1 is used where the bandwidth is chosen subjectively in order to get a suitable trade off between smoothness of the estimate and level of detail in the estimate. There is no obvious trend in the estimated intensity surface, so we assume that the point pattern is a partially observed realization of a stationary point process.

A non-parametric estimate of $L(r) - r$, $r > 0$, (see Section 4.5.2), the estimates $\hat{G}$ (4.16) and $\hat{F}$ (4.15), the estimate of J obtained from $\hat{G}$ and $\hat{F}$, and a non-parametric estimate of the pair correlation are also shown in Figure 4.3. The pair correlation function is estimated using (4.13) with $\widehat{\rho(\xi)\rho(\eta)}$ given by (4.10), and an *Epanecnikov kernel*

$$k_2(r) = 1[|r| < b]3(1 - (r/b)^2)/(4b)$$

with bandwidth $b = 2$. For each summary statistic, 2.5% and 97.5% envelopes are calculated from 39 simulations of the fitted homogeneous Poisson process with intensity $\hat{\rho} = 134/(56 \times 38)$.

There is clear indication of regularity since the estimate of $L(r) - r$ is smaller than 0 and below the lower envelope for r up to around 8. Similarly, $\hat{G}$ is smaller than expected for a Poisson process, and $\hat{F}(r)$ is above the upper envelope for $r > 2$. Also the estimated J-function and pair correlation function suggest that the point pattern is regular. Note from the envelopes that the pair correlation estimate appears to be biased upwards for $0 < r < 1$ under the fitted Poisson model. So it does not seem advisable to interpret the small kink of the estimated $g(r)$ occurring for $0 < r < 1$.

4.5.5 Example 1: Weed plants (continued)

Brix & Møller (2001) observe a log linear trend for the intensity of the weed plants perpendicular to the ploughing direction, so they consider a parametric log linear model for the intensity function,

$$\log \rho(\xi; \theta) = \theta_1 + \theta_2 \xi_2, \quad \xi = (\xi_1, \xi_2) \in W,\ \theta = (\theta_1, \theta_2) \in \mathbb{R}^2, \tag{4.17}$$

where W is the union of the 45 observation frames. This is supported by the fact that the humidity of the field seemed to have a gradient in that

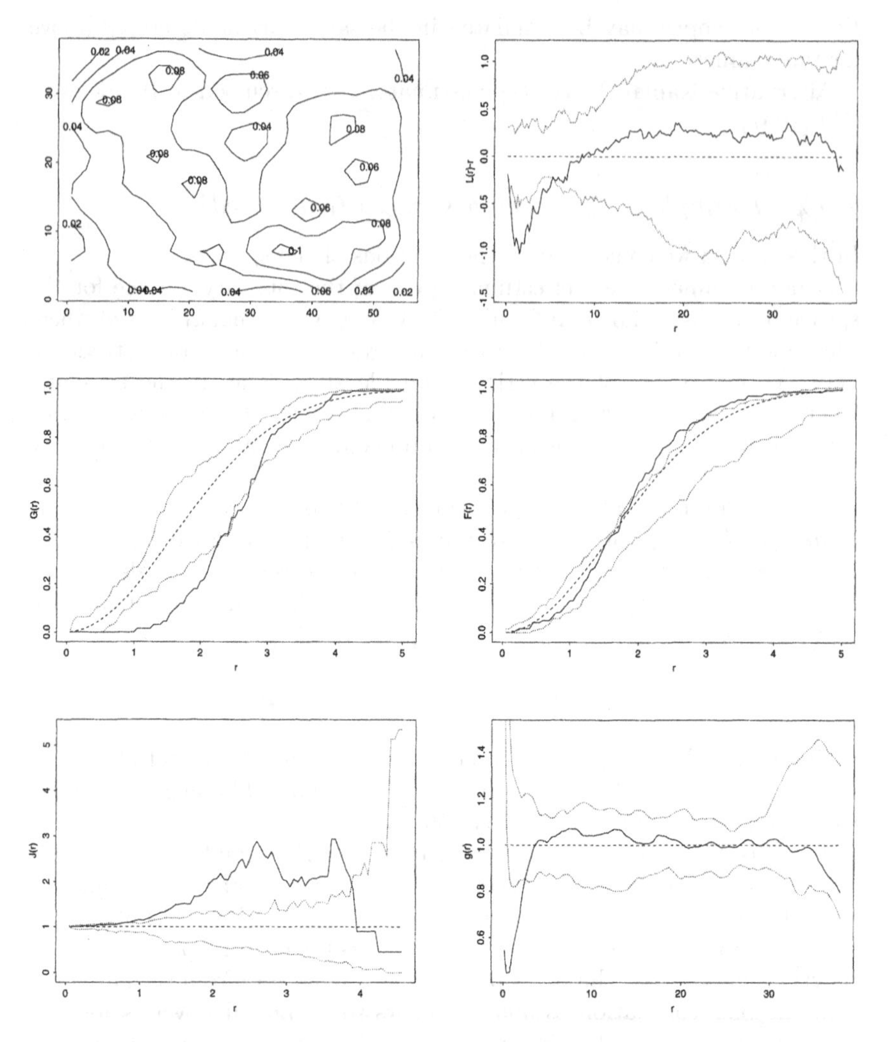

FIGURE 4.3. Summary statistics for positions of Norwegian spruces. Upper left: estimate of intensity surface. Upper right: estimated $L(r) - r$ (solid line) and envelopes calculated from 39 simulations under the fitted homogeneous Poisson process. Dashed line is the theoretical value of $L(r) - r$ for a Poisson process. Middle left, middle right, lower left, and lower right: as upper right but for G, F, J, and g.

direction; a fact that should stimulate the occurrence of *Trifolium* spp. In Brix & Møller (2001) the parameter θ is estimated in an ad hoc manner using a simple linear regression with dependent variables given by the weed counts in each frame.

Alternatively we here consider the log likelihood under the assumption that the weed plants form a Poisson process. By (4.3) the log likelihood is given by

$$\sum_{\eta \in x_W} \log \rho(\eta;\theta) - \log \int_W \rho(\xi;\theta) \mathrm{d}\xi \tag{4.18}$$

where x_W is the set of weed plant locations. As suggested in Berman & Turner (1992) and further investigated in Baddeley & Turner (2000), likelihoods of the form (4.18) can easily be maximised using standard software for generalised linear models, see also Section 4.7.2. Using this approach with the Splus routine `glm()` we obtain the maximum likelihood estimate $\hat{\theta} = (-4.10, 0.003)$ (which is close to the estimate $(-4.28, 0.003)$ obtained in Brix & Møller (2001)).

By replacing $\bar{\rho}(\cdot)$ in (4.11) with $\rho(\cdot;\hat{\theta})$ an estimate of K_{inhom} and thereby of L_{inhom} is obtained; denote this estimate by $\hat{L}_{\text{inhom},\hat{\theta}}$. Figure 4.4 shows $\hat{L}_{\text{inhom},\hat{\theta}}(r) - r$ which should be close to zero if the weed positions were Poisson. The dashed line shows the average of estimated $L_{\text{inhom},\hat{\theta}}(r) - r$ functions calculated from 39 simulations under the fitted inhomogeneous Poisson process (i.e. when the intensity function is assumed to be known and given by $\rho(\cdot;\hat{\theta})$). It appears that $\hat{L}_{\text{inhom},\hat{\theta}}$ is nearly unbiased under the fitted Poisson process. Furthermore, the upper/lower envelopes calculated from the 39 simulations of the fitted inhomogeneous Poisson process are rather constant when $r \leq 20$ cm, whilst they increase/decrease for larger values of r. This may be compared with the conditions (4.8) and (4.9) which require r to be less than 20 cm and $(30^2 + 60^2)^{1/2} \approx 67$ cm, respectively. Arguably, the envelopes may be too narrow since we are ignoring the variability of $\hat{\theta}$ when using the fitted Poisson model for the simulations.

A plot similar to that of the L-function but for the pair correlation function g is also shown in Figure 4.4. Here g is estimated by (4.13) with an Epanecnikov kernel with bandwidth 3 and $\hat{\rho}(\cdot)$ given by $\rho(\cdot;\hat{\theta})$. As in Section 4.5.4 the pair correlation function estimate is biased upwards for small distances under the fitted Poisson process.

By Figure 4.4, the weed data clearly exhibit clustering, since $\hat{L}_{\text{inhom},\hat{\theta}}(r) - r$ takes positive values above the upper envelope when $r \leq 20$ cm. Similarly, the estimated $g(r)$ falls above the upper envelope for distances up to around 17. Note that the envelopes for $g(r)$ differ most for $20 < r < 40$ where few interpoint distances are observed due to the experimental design.

A close look at Figure 4.1 shows that the intensities of weed plants are higher in the third column of frames than in the first column of frames. This is not reflected by the log linear model (4.17), so an appropriate alternative might be a log third order polynomial model

$$\log \rho_3(\xi;\psi) = \psi_1 + \psi_2\xi_2 + \psi_3\xi_2^2 + \psi_4\xi_2^3, \tag{4.19}$$

where $\xi = (\xi_1, \xi_2) \in W$ and $\psi = (\psi_1, \psi_2, \psi_3, \psi_4) \in \mathbb{R}^4$. The maximum likelihood estimate is

$$\hat{\psi} = (-3.670, -0.014, 9.342 \times 10^{-5}, -1.310 \times 10^{-7}).$$

Using $\rho_3(\cdot;\hat{\psi})$ in (4.11) we obtain an estimate $\hat{L}_{\text{inhom},3,\hat{\psi}}(r) - r$. The plot (omitted) of $\hat{L}_{\text{inhom},3,\hat{\psi}}(r) - r$ is qualitatively similar to the plot of the estimated $L(r) - r$ function in Figure 4.4: the clustering is less pronounced, but the Poisson model is still rejected.

As a more appropriate model, Brix & Møller (2001) consider a log Gaussian Cox process (see Section 4.6.2) where it is possible to model clustering due to environmental effects and the seed bank in the soil. In Section 4.6.3 we also consider a log Gaussian Cox process model for the weed plants.

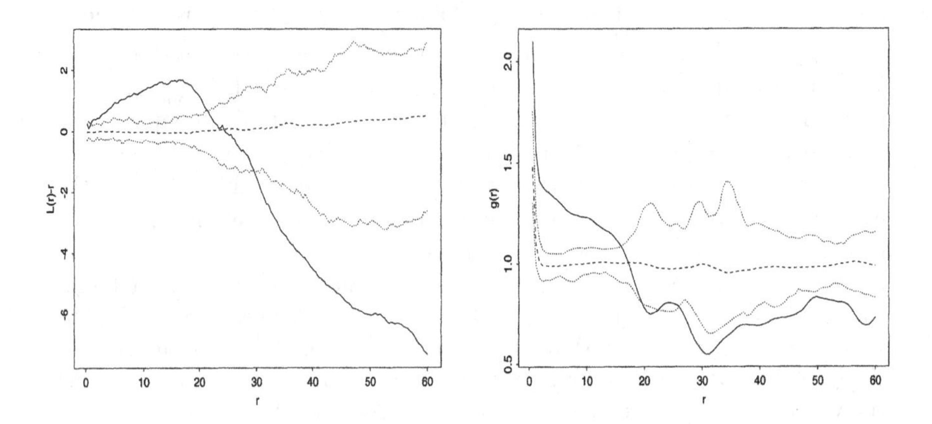

FIGURE 4.4. Left plot: solid line is $\hat{L}_{\text{inhom},\hat{\theta}}(r) - r$ for weed plants; dashed horizontal line is average of $L_{\text{inhom},\hat{\theta}}(r) - r$ functions computed from 39 simulations under the fitted Poisson model; dotted lines are 2.5% and 97.5% envelopes for $\hat{L}_{\text{inhom},\hat{\theta}}(r) - r$ obtained from the 39 simulations. Right plot: as left plot, but for the pair correlation function.

4.6 Models and simulation-based inference for aggregated point patterns

Aggregation in a spatial point process may be caused by at least three factors: (i) spatial heterogeneity, e.g. due to some underlying "environmental" process, (ii) clustering of the points around the points of another point process, (iii) interaction between the points. In this section we concentrate mostly on (i) and partly on (ii), while (iii) is considered in Section 4.7.1. The relationship between (i) and (ii) is described in Section 4.6.1 which deals with Cox and cluster processes; further material on such processes can be found in Diggle (1983), Stoyan et al. (1995), and the references therein. In Section 4.6.2 we study the particular case of log Gaussian Cox processes. Other specific models such as shot noise Cox processes are discussed in Section 4.6.4.

4.6.1 Cox and cluster processes

Definitions and properties of Cox processes

A natural way to generalise the definition of a Poisson point process is to let Λ be a random, diffuse, and locally finite measure on S so that conditional on Λ, $X|\Lambda \sim \text{Poisson}(S, \Lambda)$. Thereby we obtain a *doubly stochastic Poisson point process* which is also called a *Cox process* with *driving measure* Λ. Specific constructions of Λ are considered in Sections 4.6.2 and 4.6.4, and in the following two simple examples.

The simplest non-trivial example of a Cox process is a *mixed Poisson process* in $\mathbb{R}^d$. This is obtained by letting $R > 0$ be a random variable and $X|R$ a homogeneous Poisson process in $\mathbb{R}^d$ with intensity R. For example, if R is gamma distributed, $X(A)$ follows a negative binomial distribution for $A \in \mathcal{B}_0$.

Another example is *random independent thinning of a Poisson process:* Suppose that conditional on a random field $\Pi = \{\Pi(\xi) : \xi \in S\}$ with $0 \leq \Pi(\cdot) \leq 1$, $X \sim \text{Poisson}(S, \mu)$ and $R(\xi) \sim \text{Uniform}[0, 1]$, $\xi \in S$, are mutually independent. Then $Z = \{\xi \in X : R(\xi) < \Pi(\xi)\}$ is a Cox process driven by the random measure given by $\Lambda(B) = \int_B \Pi(\xi)\mu(\mathrm{d}\xi)$ (provided this integral is well-defined for Borel sets $B \subseteq S$, and Λ is locally finite and diffuse).

By definition of a Cox process X and the properties of Poisson processes we obtain immediately the following general results. The void probabilities are given by

$$P(X(B) = 0) = \mathbb{E}\exp(-\Lambda(B)).$$

Furthermore, the intensity measure is given by

$$\mu(B) = \mathbb{E}X(B) = \mathbb{E}\Lambda(B),$$

and the second order factorial moment measure by

$$\alpha^{(2)}(A \times B) = \mathbb{E} \sum_{\xi,\eta \in X}^{\neq} 1[\xi \in A, \eta \in B] = \mathbb{E}[\Lambda(A)\Lambda(B)].$$

Combining these results, we see that

$$\mathbb{V}\text{ar}(X(B)) = \mathbb{V}\text{ar}(\Lambda(B)) + \mu(B).$$

Thus, compared to a Poisson process, a Cox process exhibits overdispersion.

In the particular case of $S = \mathbb{R}^d$, if Λ has a (random) density $\mathrm{d}\Lambda(\xi)/\mathrm{d}\xi = \lambda(\xi)$, then X has intensity function

$$\rho(\xi) = \mathbb{E}\lambda(\xi) \tag{4.20}$$

and second order product density

$$\rho^{(2)}(\xi, \eta) = \mathbb{E}[\lambda(\xi)\lambda(\eta)]. \tag{4.21}$$

Closed form expressions of these functions can sometimes be derived for specific models, cf. Sections 4.6.2 and 4.6.4.

Definitions and properties of cluster processes

Let Y be a point process defined on a space T, and for each $\xi \in Y$, let Z_ξ be a point process defined on a space S so that the superposition

$$X = \bigcup_{\kappa \in Y} Z_\kappa$$

is a simple and locally finite point process on S. Then X is called a *cluster process* with *mother process* Y and *clusters* or *daughters* Z_κ, $\kappa \in Y$. We may e.g. think of plants (the mothers) that spread seeds (the daughters). Usually in applications, $S = T \subseteq \mathbb{R}^d$ and the mother process is unobserved, so we are dealing with a missing data problem.

Certain cluster processes are special cases of Cox processes: If conditional on Y, the clusters are independent, each cluster $Z_\kappa \sim \text{Poisson}(S, \mu_\kappa)$, and the random measure given by

$$\Lambda(B) = \sum_{\kappa \in Y} \mu_\kappa(B), \quad B \subseteq S,$$

is locally finite and diffuse, then X is a Cox process driven by Λ. This follows simply by finding the void probabilities of X.

A particular important subclass of such models are *Neyman-Scott processes* X, where it is assumed that $S = \mathbb{R}^d$, each cluster Z_κ has an intensity function $\rho_\kappa(\xi)$, and $Y \sim \text{Poisson}(T, \mu)$ (slightly extending the definition of Neyman & Scott (1958)). Then Λ has density

$$\lambda(\xi) = \sum_{\kappa \in Y} \rho_\kappa(\xi)$$

with respect to Lebesgue measure, which combined with (4.20) and (4.21) give expressions for the intensity function and second order product density function of X,

$$\rho(\xi) = \mathbb{E}\sum_{\kappa \in Y} \rho_\kappa(\xi), \quad \rho^{(2)}(\xi, \eta) = \mathbb{E}\sum_{\kappa, \zeta \in Y} \rho_\kappa(\xi)\rho_\zeta(\eta).$$

Then by Slivnyak-Mecke (4.1)-(4.2),

$$\rho(\xi) = \int \rho_\kappa(\xi) \mathrm{d}\mu(\kappa)$$

and

$$\rho^{(2)}(\xi, \eta) = \int \int \rho_\kappa(\xi)\rho_\zeta(\eta) \mathrm{d}\mu(\kappa)\mathrm{d}\mu(\zeta) + \int \rho_\kappa(\xi)\rho_\kappa(\eta) \mathrm{d}\mu(\kappa).$$

These expressions can be further reduced when $T = \mathbb{R}^d$, Y is a homogeneous Poisson process, and $Z_\kappa - \kappa$, $\kappa \in Y$, (the clusters relative to their mother points) are i.i.d. and independent of Y; see e.g. Stoyan et al. (1995).

However, closed form expressions of $\rho^{(2)}$ and hence the pair correlation can only be derived for a few such Neyman-Scott processes, including a *Thomas process* X defined as follows: Y is a homogeneous Poisson process with intensity $\rho_Y > 0$, and $Z_\kappa - \kappa$ is a Poisson process where the number of points is po(α)-distributed and each point follows a d-dimensional normal distribution with mean 0 and covariance matrix $\sigma^2 I$. Then X is stationary and isotropic with intensity $\rho_X = \alpha\rho_Y$ and pair correlation function $g(\xi, \eta) = g(\|\xi - \eta\|)$ given by

$$g(r) = 1 + \exp(-r^2/(4\sigma^2))/[\lambda_Y(4\pi\sigma^2)^{d/2}], \quad r \geq 0.$$

Furthermore, in the planar case $d = 2$, Ripley's K-function is

$$K(r) = \pi r^2 + [1 - \exp(-r^2/(4\sigma^2))]/\lambda_Y.$$

Simulation procedures for Neyman-Scott processes on a bounded window B follow often straightforwardly from their definition as cluster processes or their construction as Cox processes: in either case, we first simulate Y and next $X_B|Y$. In order to avoid boundary effects the mother process usually must be simulated on an extended window $A \supset B$ so that daughter points from a mother outside A falls into B with a negligible probability; see Brix & Kendall (2002) and Møller (2002b).

4.6.2 Log Gaussian Cox processes

Definitions and properties

Suppose that $Y = \{Y(\xi) : \xi \in \mathbb{R}^d\}$ is a real-valued Gaussian process, i.e. any finite linear combination of the $Y(\xi)$ follows a normal distribution. If X is a Cox process on $\mathbb{R}^d$ driven by a random measure with density

$$\lambda(\xi) = \exp(Y(\xi))$$

with respect to Lebesgue measure, then X is said to be a *log Gaussian Cox process (LGCP)*. Such models have independently been introduced in astronomy by Coles & Jones (1991) and in statistics by Møller et al. (1998).

It is necessary to impose weak conditions on the mean function $m(\xi) = \mathbb{E}Y(\xi)$ and covariance function $c(\xi, \eta) = \mathbb{C}\mathrm{ov}(Y(\xi), Y(\eta))$ in order to get a well-defined and finite integral $\int_B \exp(Y(\xi))\mathrm{d}\xi$ for bounded Borel sets $B \subset \mathbb{R}^d$. For example, we may require that $\xi \to Y(\xi)$ is almost surely continuous. This is the case if m and c are continuous functions, and for some $0 < C < \infty$ and some $\epsilon > 0$,

$$c(\xi, \xi) + c(\eta, \eta) - 2c(\xi, \eta) \leq C/(-\log \|\xi - \eta\|)^{1+\epsilon} \tag{4.22}$$

whenever $\|\xi - \eta\| < 1$, cf. Theorem 3.4.1 in Adler (1981). These are fairly weak conditions which are usually satisfied for the models used in practice.

The definition of an LGCP can easily be extended in a natural way to multivariate LGCPs as shown in Møller et al. (1998) and to multivariate spatio-temporal LGCP as studied in Brix & Møller (2001). LGCPs are flexible models for clustering as demonstrated in Møller et al. (1998), where examples of covariance functions together with simulated realizations of LGCPs and their underlying Gaussian processes are shown. Certain Thomas processes may in practice be difficult to distinguish from LGCPs with a Gaussian covariance function $c(\xi, \eta) = \sigma^2 \exp(-(\|\xi - \eta\|/\alpha)^2)$, where $\sigma^2 > 0$ and $\alpha > 0$ are parameters (Møller et al. 1998).

The intensity function and pair correlation function of an LGCP are simply given by

$$\rho(\xi) = \exp(m(\xi) + c(\xi, \xi)/2), \quad g(\xi, \eta) = \exp(c(\xi, \eta)). \tag{4.23}$$

Hence, if $c(\xi, \eta) = c(\xi - \eta)$ is translation invariant, K_{inhom} exists. Further theoretical results for the intensities of an LGCP (including third and higher order properties) can be found in Møller et al. (1998). These results are in general of a different and much simpler form than for other Cox processes.

By (4.23) there is a one-to-one correspondence between c and g and between (m, c) and (ρ, g). Consequently, the distribution of an LGCP is uniquely determined by its first- and second-order properties as given by the intensity and pair correlation functions. This makes parametric models easy to interpret and simple methods for parameter estimation and model checking become available as discussed in Møller et al. (1998).

Finally, notice that there is no problem with edge effects as the distribution of an LGCP restricted to a bounded subset is known.

Simulation of LGCPs

Unconditional simulation: Below we describe shortly how to simulate an LGCP X and its underlying Gaussian process Y when both are restricted to a bounded region B. Since $X_B|Y$ is simply a Poisson process with intensity function $\exp(Y)$ on B, we restrict attention to simulation from $Y_B = \{Y(\xi) : \xi \in B\}$. As the infinitely dimensional process Y_B does not in general have a finite representation in a computer, we approximate Y_B by a random step function with constant value $Y(c_i)$ within disjoint cells C_i, where $B = \cup_{i \in I} C_i$, I is a finite index set, and $c_i \in C_i$ is a "centre" point of C_i. So we actually consider how to simulate the Gaussian vector $\tilde{Y} = (\tilde{Y}_i)_{i \in I}$ where $\tilde{Y}_i = Y(c_i)$.

Suppose for the moment that B is rectangular, say $B = [0, 1[^2$, and let $I \subset B$ denote a regular grid. As discussed in Møller et al. (1998), there is an efficient way of simulating $\tilde{Y}$ when $c(\xi, \eta) = c(\xi - \eta)$ is invariant under translations. Briefly, I is embedded in a rectangular grid I_{ext}, which is wrapped on a torus, and a block circulant matrix $K = \{K_{ij}\}_{i,j \in I_{\text{ext}}}$ is constructed so that the submatrix $\{K_{ij}\}_{(i,j) \in I}$ is the covariance matrix of $\tilde{Y}$. Since K is block circulant, it can easily be diagonalised by means of the two-dimensional discrete Fourier transform with associated matrix F_2 (see Section 6.1 in Møller et al. (1998) and Wood & Chan (1994)). Suppose that K is positive semi-definite (i.e. K has non-negative eigenvalues). Then we can extend $\tilde{Y} = (\tilde{Y}_{(i,j)})_{(i,j) \in I}$ to a larger Gaussian field $\tilde{Y}_{\text{ext}} = (\tilde{Y}_{(i,j)})_{(i,j) \in I_{\text{ext}}}$ with covariance matrix K: set

$$\tilde{Y}_{\text{ext}} = \Gamma Q + \mu_{\text{ext}} \tag{4.24}$$

where Γ follows a standard multivariate normal distribution, the restriction of μ_{ext} to I agrees with the mean of $\tilde{Y}$, and $Q = \bar{F}_2 \tilde{\Lambda}^{1/2} F_2$, where $\tilde{\Lambda}$ is the diagonal matrix of eigenvalues for K, and $\bar{F}_2$ is the complex conjugate of F_2. Using the two-dimensional *fast Fourier transform* a fast simulation algorithm for $\tilde{Y}_{\text{ext}}$ and hence $\tilde{Y}$ is obtained.

Another possibility is to use the *Cholesky decomposition* of the covariance matrix of $\tilde{Y}$, provided this covariance matrix is positive definite. This may be advantageous if c is not translation invariant or B is far from being rectangular, see Section 4.6.3. On the other hand, the Cholesky decomposition is only practically applicable if the dimension of $\tilde{Y}$ is moderate. We can still refer to (4.24) when the Cholesky decomposition is used, letting now I_{ext}, $\tilde{Y}_{\text{ext}}$, K, and Q be specified as follows: $I_{\text{ext}} = I$, $\tilde{Y}_{\text{ext}} = \tilde{Y}$, K is the covariance matrix of $\tilde{Y}$, and Q denotes the upper-triangular matrix obtained from the Cholesky decomposition.

Conditional simulation: Now, suppose that we have observed a point pattern $X_W = x$ within a window $W \subseteq B$. When making conditional simulations of (Y_B, X_A) given X_W, we may simulate first from $Y_B | X_W$ and next from $X_A | (Y_B, X_W)$. The latter conditional distribution is just a Poisson process with intensity function $\exp(Y_A)$ on A, so we restrict attention to simulation of $Y_B | X_W$ below. Note that if we wish to make further conditional simulations within a region D which is disjoint to B, we may first make simulations from $Y_D | (Y_B, X_B)$, which is Gaussian and does not depend on X_B, and next from the Poisson process $X_D | Y_D$ with intensity function $\exp(Y_D)$ on D.

Approximate simulations of $Y_B|X_W = x$ can be obtained from simulations of $\tilde{Y}|X_W = x$ which in turn can be obtained from simulations of $\Gamma|X_W = x$ using the transformation (4.24). Omitting an additive constant depending on x only, the log conditional density of Γ given x is

$$-\|\gamma\|^2/2 + \sum_{i \in I} (\tilde{y}_i n_i - A_i \exp(\tilde{y}_i)) \tag{4.25}$$

where, in accordance with (4.24), $(\tilde{y}_i)_{i \in I_{\text{ext}}} = \gamma Q + \mu_{\text{ext}}$, $n_i = x(C_i)$, and $A_i = |C_i \cap W|$. Note that (4.25) is formally equivalent to the conditional density of the random effects given observations n_i, $i \in I$, in a generalised linear mixed model with Poisson error distribution, cf. Section 2.9.1. The gradient of (4.25) becomes

$$\nabla(\gamma) = -\gamma + \left(n_i - A_i \exp(\tilde{y}_i)\right)_{i \in I_{\text{ext}}} Q^{\mathsf{T}},$$

and differentiating once more the conditional density of Γ given x is seen to be strictly log-concave.

For simulation from $\Gamma|X_W = x$, Møller et al. (1998) use a *Langevin-Hastings algorithm* or *Metropolis-adjusted Langevin algorithm* as introduced in the statistical community by Besag (1994) (see also Roberts & Tweedie (1996) and Section 1.3.1) and earlier in the physics literature by Rossky, Doll & Friedman (1978). This is a Metropolis-Hastings algorithm with collective updating inspired by the definition of a Langevin diffusion. If γ is the current state generated by the Langevin-Hastings algorithm, then we first propose a new state generated from a multivariate normal distribution with mean $\gamma + (h/2)\nabla(\gamma)$ and covariance matrix hI_d, where I_d is the $d \times d$ identity matrix and $h > 0$ is a user-specified parameter. Secondly we accept or reject the proposal in accordance to the Hastings ratio. Theoretical results in Roberts & Rosenthal (1998) and Breyer & Roberts (2000) suggest that one should tune h to obtain acceptance rates around 0.57. The use of the gradient in the proposal distribution may lead to much better convergence properties when compared to the standard alternative of a random walk Metropolis algorithm, see Christensen, Møller & Waagepetersen (2001) and Christensen & Waagepetersen (2002). See also Sections 1.3.1 and 2.9.1.

A truncated version of the Langevin-Hastings algorithm is obtained by replacing the gradient $\nabla(\gamma)$ by

$$\nabla_{\text{trun}}(\gamma) = -\gamma + \left(n_i - \min\{H, A_i \exp(\tilde{y}_i)\}\right)_{i \in I_{\text{ext}}} Q^{\mathsf{T}} \tag{4.26}$$

where $H > 0$ is a user-specified parameter which can e.g. be taken to be twice the maximal n_i, $i \in I$. We still tune h so that the acceptance rate is about 0.57. As shown in Møller et al. (1998) the *truncated Langevin-Hastings algorithm* is geometrically ergodic.

The algorithms described above can easily be generalised to the case of multivariate LGCPs, cf. Møller et al. (1998).

Bayesian inference for LGCPs

Møller et al. (1998) consider an empirical Bayesian approach to *prediction* of Y_B given $X_W = x$, where a parametric model has been chosen for the mean and covariance functions and the parameters are estimated by a so-called minimum contrast method. This estimation method depends on certain user-specified parameters, and the uncertainty of the estimated parameters are not taken into account, so arguably the variation of $Y_B|X_W = x$ may be underestimated. In the sequel we consider an alternative *fully Bayesian approach* with hyper priors on the parameters and discuss how we can make MCMC simulations of the full posterior given $X_W = x$. Again the random step function approximation of Y_B is used, and we consider the posterior of Γ from which the posterior of $\tilde{Y}$ can be computed.

Specifically, assume that the mean function m restricted to B is a linear function $m(\xi) = \theta z(\xi)^\mathsf{T}$ of a p-dimensional covariate $z(\xi)$, and the covariance function is of the form $c(\xi, \eta) = \sigma^2 r(\|\xi - \eta\|/\alpha))$ where $\sigma > 0$ is the standard deviation of $Y(\xi)$ and $\alpha > 0$ is a scale parameter for the correlation. Such a situation is considered in Benes, Bodlak, Møller & Waagepetersen (2002) and in Section 4.6.3 below. We impose independent hyper priors p_1, p_2, and p_3 on θ, σ, and $\kappa = \log \alpha$, respectively. The posterior density is thus

$$\pi(\gamma, \theta, \kappa, \sigma | x) \propto p_1(\theta) p_2(\sigma) p_3(\kappa) \\ \times \exp\Big(- \|\gamma\|^2/2 + \sum_{i \in I} (\tilde{y}_i n_i - A_i \exp(\tilde{y}_i)) \Big), \tag{4.27}$$

cf. (4.25). Note that $\tilde{y}$ is a function of $(\gamma, \theta, \kappa, \sigma)$.

As in Christensen & Waagepetersen (2002), Christensen et al. (2001), and Benes et al. (2002) we use a hybrid algorithm (also called Metropolis-within-Gibbs, cf. Section 1.6.1) with a systematic MCMC updating scheme for the full conditional distribution of each of the four parameters $\gamma, \theta, \kappa, \sigma$. For γ we use a truncated Langevin-Hastings algorithm where the gradient by (4.27) is still given by (4.26). For θ we use also a truncated Langevin-Hastings algorithm with gradient

$$\nabla_{\text{trun}}(\theta) = \big(n_i - \min\{H, A_i \exp(\tilde{y}_i)\}\big)_{i \in I_{\text{ext}}} D + \frac{\partial}{\partial \theta} \log p_1(\theta)$$

where D is the design matrix with rows given by $z(c_i)$, $i \in I$ (as $n_i = A_i = 0$ for cells outside W, we need only to specify D for centre points contained in W). The same truncation constant H is used in the two Langevin-Hastings algorithms, and as before the variances in the proposal distributions are tuned so that acceptance rates about 0.57 are obtained. Finally, random walk Metropolis updates are used for κ and $\log \sigma$, respectively, with acceptance rates around the optimal value 0.23 (Roberts, Gelman & Gilks 1997). See also Section 1.5.3.

4.6.3 Example 1: Weed plants (continued)

For the weed data, B is the union of the 45 observation frames and we use a discretisation of the Gaussian field where each of the 45 frames is subdivided into 6 quadratic cells of side length 10 cm. The covariance matrix of the 270 dimensional vector $\tilde{Y}$ is decomposed using the Cholesky decomposition. The covariate vector is $z(\xi) = (z_1(\xi), z_2(\xi)) = (1, \xi_2)$ for $\xi = (\xi_1, \xi_2) \in B$, and we use the exponential correlation function $r(\|\xi\|) = \exp(-\|\xi\|)$. The hyper priors are chosen to be $p_1(\theta) \propto 1$, $\theta \in \mathbb{R}^2$, $p_2(\sigma) \propto \exp(-10^{-5}/\sigma)/\sigma$, $\sigma > 0$, and $p_3(\kappa) \propto 1$, $\log 0.75 < \kappa < \log 75$. The improper prior p_1 is completely flat, and the improper p_2 yields an essentially flat prior for $\log \sigma$ on $(0, \infty)$. The limits for the log uniform prior p_3 were chosen subjectively in order to accommodate a reasonable range of strengths of correlation. For the discretised LGCP one can check as in Christensen & Waagepetersen (2002) that these priors yield a proper posterior but strictly speaking we do not know whether a proper posterior is also obtained for the original LGCP. One may therefore consider the possibility of restricting the supports of θ and σ to large but bounded regions.

The hybrid algorithm described at the end of Section 4.6.2 is used for the computations. In order to improve mixing of the Markov chain we use a *reparameterisation* (see Section 1.3.2) where $(z_2(c_i))_{i \in I}$ is normalised to have zero mean and maximum absolute value equal to one. The posterior distributions shown in Figure 4.5 are computed from time series obtained by subsampling each 10th of 200,000 scans of the hybrid algorithm; here a scan means an update of each of γ, θ, σ, κ in a systematic order. Plots of the time series (omitted) suggest that equilibrium is attained after around 400 scans and according to estimated autocorrelations, the states in the time series are uncorrelated for lags greater than 30. The posterior means

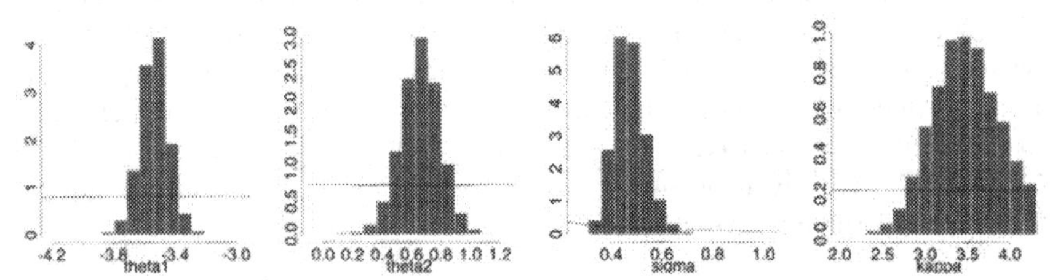

FIGURE 4.5. Posterior distributions of θ_1, θ_2, σ, and κ. The solid lines indicate the prior distributions.

of θ_1 and θ_2 are -4.11 and 0.003, i.e. very close to the maximum likelihood estimates for the inhomogeneous Poisson process considered in Section 4.5.5. The posterior means of σ and κ are 0.47 and 3.48. The value $\kappa = 3.48$ yields correlations 0.40 and 0.05 at distances 30 cm and 100 cm, respectively. Note that the posterior for κ is very sensitive to the choice of prior — one can in fact verify that the posterior support is equal to the prior support. The posterior mean of $\tilde{Y}$ is shown in Figure 4.6 where large posterior means of $\tilde{Y}_i$ coincide with cells C_i containing many weed plants.

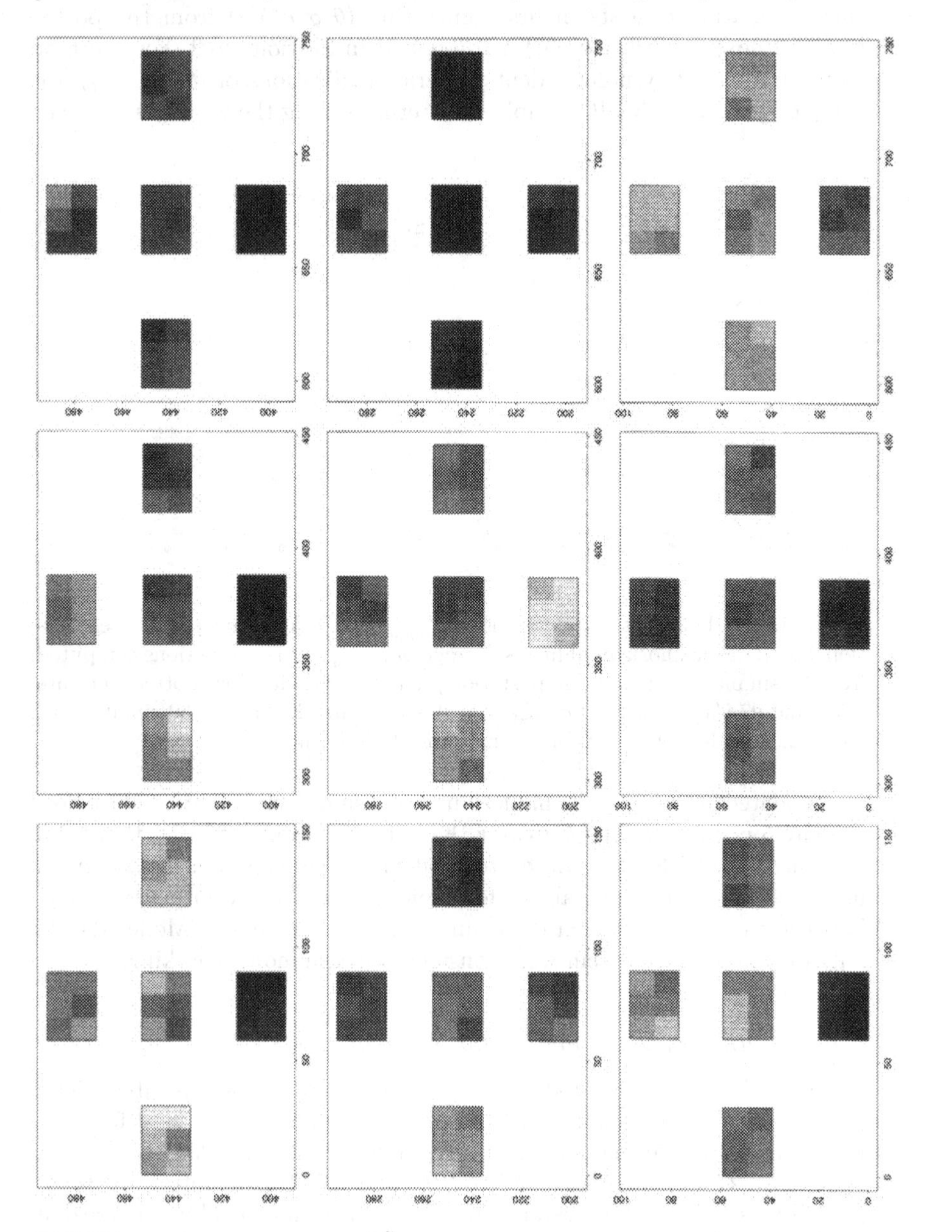

FIGURE 4.6. Posterior means of $\tilde{Y}$. Values range between −0.60 (light grey) and 0.97 (dark grey).

Figure 4.7 is similar to Figure 4.4. It shows $\hat{L}_{\text{inhom},\hat{\theta}}(r) - r$ and $\hat{g}(r)$ from Section 4.5.5 but now with envelopes calculated from simulations under the posterior predictive distribution (Gelfand 1996) for the LGCP. Each such simulation consists in first generating $(\theta, \sigma, \kappa, Y_B)$ from the posterior and next $X_W|(\theta, \sigma, \kappa, Y_B)$ as discussed in Section 4.6.2. For the first step, approximately independent posterior realizations of $(\theta, \sigma, \kappa, Y_B)$ are sampled from the MCMC sample. Clustering among the weed plants is ac-

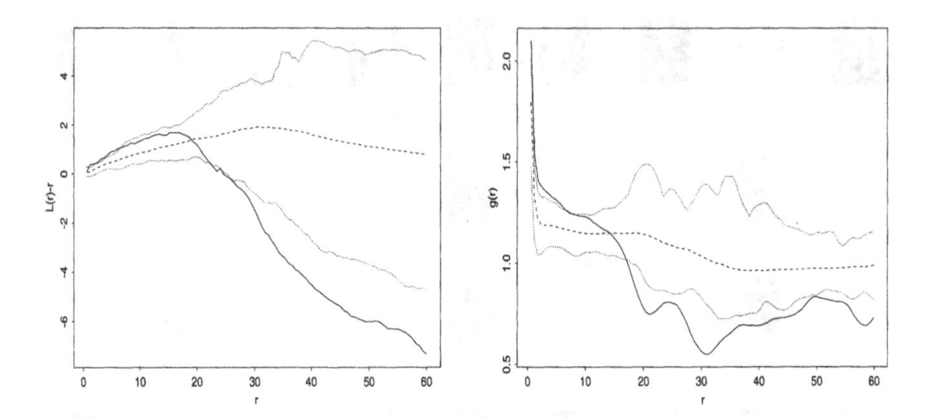

FIGURE 4.7. Left plot: solid line is $\hat{L}_{\text{inhom},\hat{\theta}}(r) - r$ for weed plants (see Section 4.5.5); dashed horizontal line is average of $L_{\text{inhom},\hat{\theta}}(r) - r$ functions computed from 39 simulations under the posterior predictive distribution; dotted lines are 2.5% and 97.5% envelopes for $\hat{L}_{\text{inhom},\hat{\theta}}(r) - r$ obtained from the 39 simulations. Right plot: as left plot, but for the pair correlation function.

commodated by the LGCP model but both summary statistics fall below the envelopes for distances greater than 20 cm. As noted in Section 4.5.5 the linear model for m may be too inflexible. Perhaps it is necessary to use a model for m which allows for rapid changes in the intensity as e.g. between the second and third row in Figure 4.1. In Brix & Møller (2001) only distances up to 20 cm were considered in the model checking.

4.6.4 Other specific models for Cox processes

So far we have concentrated much on LGCPs. Below we describe briefly two other interesting classes of Cox processes which can be used for non-parametric Bayesian modelling: the Heikkinen & Arjas (1998) model and *shot noise G Cox processes (SNGCP)* (Brix 1999). We introduce these models by specifying their random intensity function λ on a bounded region $R \subseteq \mathbb{R}^2$.

In Heikkinen & Arjas (1998) $\lambda(\xi) = \sum_k \lambda_k \mathbf{1}_{A_k}(\xi)$ where $\{A_k\}$ is the Voronoi tessellation generated by a point process of nuclei $\{y_k\} \subset R$, i.e. A_k is the set of points in R closer to y_k than to any other nuclei. The nuclei follow a homogeneous Poisson process restricted to R, and conditional on

$\{y_k\}$, $\{\log \lambda_k\}$ is modelled by a conditional autoregression (Besag 1974).

The construction of a SNGCP is a bit more complicated. Now, $\lambda(\xi) = \sum_j k(\xi, u_j)\gamma_j$ where k is a kernel (for simplicity we assume that $k(\cdot, u)$ is a density function for a continuous random variable), and $\{(u_j, \gamma_j)\} \subset E \times [0, \infty)$ where E is a given planar region (a more general setting is considered in Wolpert & Ickstadt, 1998). Typically in applications, $E = R$, or in order to reduce edge effects, $R \subset E$ where E is much larger than R. Further, $\{(u_j, \gamma_j)\}$ is a Poisson process with intensity measure

$$\nu(A \times B) = (\alpha(A)/\Gamma(1-\kappa)) \times \int_B \gamma^{-\kappa-1} \exp(-\tau\gamma) \mathrm{d}\gamma, \quad A \subseteq E,\ B \subseteq [0, \infty),$$

where $\kappa < 1$ and $\tau \geq 0$ are parameters with $\tau > 0$ if $\kappa \leq 0$, and α is a locally finite measure (or, as in Brix, 1999, a nonnegative and nonzero Radon measure). If $\kappa < 0$, we obtain a kind of modified Neyman-Scott process as $\{u_j\}$ is a Poisson process with intensity measure $(\tau/|\kappa|)\alpha$, and $\{u_j\}$ is independent of the "marks" $\{\gamma_j\}$, which in turn are mutually independent and follow a common Gamma distribution $\Gamma(|\kappa|, \tau)$ (in a usual Neyman-Scott process, all marks are equal and deterministic). The situation is less simple for $\kappa \geq 0$ as $\{u_j\}$ is not locally finite. For $\kappa = 0$, we have a Poisson/gamma model (Daley & Vere-Jones 1988, Wolpert & Ickstadt 1998). As noticed in Wolpert & Ickstadt (1998) we may extend the model by replacing the parameter τ with a positive function $\tau(u), u \in E$, and redefining

$$\nu(A \times B) = (1/\Gamma(1 - \kappa)) \int_A \int_B \gamma^{-\kappa-1} \exp(-\tau(u)\gamma)\alpha(\mathrm{d}u)\mathrm{d}\gamma.$$

The models are reviewed and compared with LGCPs in Møller (2002a) and Møller & Waagepetersen (2002). A general class of shot noise Cox processes, including SNGCPs, is studied in Møller (2002b).

4.7 Models and simulation-based inference for Markov point processes

Markov or Gibbs point processes arose in statistical physics for the description of large interacting particle systems, see e.g. Ruelle (1969), Georgii (1976), Preston (1976), and Nguyen & Zessin (1979). Van Lieshout (2000) provides a recent account of the state of the art of Markov point processes in spatial statistics; see also the reviews in Ripley (1977), Baddeley & Møller (1989), and Stoyan et al. (1995). In Section 4.7.1 we concentrate on the case of a finite point process specified by a density with respect to a Poisson point process so that a local Markov property is satisfied. Pseudo likelihood estimation for Markov processes is considered in Sections 4.7.2–4.7.3, and maximum likelihood inference based on MCMC in Sections 4.7.4–4.7.5. Bayesian analysis is surveyed in Section 4.7.6. Simulation procedures for Markov point processes using Metropolis-Hastings algorithms and spatial birth-death processes are considered in Sections 4.7.7–4.7.8, while Section 4.7.9 concerns perfect simulation.

4.7.1 Definitions and properties

In the remaining part of this chapter we consider a finite point process X which is absolutely continuous with respect to Poisson(S, μ) where $\mu(S) < \infty$. Its density is denoted f. The point process may be extended to a larger region, and f may depend on points outside S, for example in order to take care of edge effects. However, we usually suppress such dependences and write $f(x)$ for $x \in N_{\mathrm{f}}$, where $N_{\mathrm{f}} = \{x \subseteq S : n(x) < \infty\}$ denotes the set of finite point configurations in S. We equip N_{f} with the σ-algebra $\mathcal{N}_{\mathrm{f}} = \{F \in \mathcal{N} : F \subseteq N_{\mathrm{f}}\}$. So by definition of a finite Poisson process, for events $F \in \mathcal{N}_{\mathrm{f}}$,

$$P(X \in F) = \sum_{n=0}^{\infty} \exp(-\mu(S))/n! \int \cdots \int 1[\{x_1, \dots, x_n\} \in F] \\ f(\{x_1, \dots, x_n\})\mu(\mathrm{d}x_1) \cdots \mu(\mathrm{d}x_n) \quad (4.28)$$

where the term for $n = 0$ is read as $\exp(-\mu(S))1[\emptyset \in F]f(\emptyset)$. An example of such a density is given by the density (4.3) for a Poisson process, but in the following we shall construct much more interesting models exhibiting interactions between the points. Often f is assumed to be *hereditary*, that is,

$$f(x) > 0 \Rightarrow f(y) > 0 \quad \text{for } y \subset x. \quad (4.29)$$

This amounts to the positivity condition in the Hammersley-Clifford theorem for Markov random fields, see e.g. Besag (1974) and Section 3.2.

In many applications we have a *pairwise interaction point process*,

$$f(x) \propto \prod_{\xi \in x} \phi(\xi) \prod_{\{\xi, \eta\} \subseteq x} \phi(\{\xi, \eta\}) \quad (4.30)$$

where ϕ is a non-negative function for which the right hand side is integrable with respect to Poisson(S, μ). A standard example is the *Strauss process* (Strauss 1975), where

$$\phi(\xi) = \beta \text{ and } \phi(\{\xi, \eta\}) = \gamma^{1[d(\xi,\eta) \leq R]}, \quad (4.31)$$

setting $0^0 = 1$. Here $\beta > 0$, $0 \leq \gamma \leq 1$, and $R > 0$ are parameters (if $\gamma > 1$ we do not in general have integrability, cf. Kelly & Ripley, 1976). If $\gamma = 1$ we obtain $X \sim$ Poisson$(S, \beta\mu)$, while for $\gamma < 1$ there is repulsion between R-close pairs of points in X. The special case where $\gamma = 0$ is called a *hard core point process* with *hard core* R as balls of diameter R and with centres at the points in X are not allowed to overlap. The Strauss process can be extended to a *multiscale point process* (Penttinen 1984) where

$$\phi(\{\xi,\eta\}) = \gamma_i \quad \text{if } R_{i-1} < d(\xi,\eta) \leq R_i \tag{4.32}$$

with $R_0 = 0 < R_1 < \ldots < R_k = \infty$, $\gamma_k = 1$, and $k \geq 2$; integrability is ensured if $\gamma_1 = 0$ and $\gamma_2 \geq 0, \ldots, \gamma_{k-1} \geq 0$, or if $0 < \gamma_1 \leq 1$ and $0 \leq \gamma_2 \leq 1, \ldots, 0 \leq \gamma_{k-1} \leq 1$. A collection of other examples of pairwise interaction point processes can be found in van Lieshout (2000).

A fundamental characteristic is the *Papangelou conditional intensity* defined by

$$\lambda^*(x,\xi) = f(x \cup \xi)/f(x), \quad x \in N_{\mathrm{f}},\ \xi \in S \setminus x, \tag{4.33}$$

taking $a/0 = 0$ for $a \geq 0$ (Kallenberg 1984). For example, for the pairwise interaction point process (4.30),

$$\lambda^*(x,\xi) = \phi(\xi) \prod_{\eta \in x} \phi(\{\xi,\eta\}).$$

Heuristically, $\lambda^*(x,\xi)\mu(\mathrm{d}\xi)$ can be interpreted as the conditional probability of X having a point in an infinitesimal region containing ξ and of size $\mu(\mathrm{d}\xi)$ given the rest of X is x. If f is hereditary, then there is a one-to-one correspondence between f and λ^*. Often in applications $f \propto h$ is only specified up to proportionality, but $\lambda^*(x,\xi) = h(x \cup \xi)/h(x)$ does not depend on the normalising constant.

Integrability of a given function h with respect to Poisson(S,μ) may be implied by stability conditions in terms of λ^*. *Local stability* means that λ^* is uniformly bounded and f is hereditary; this implies integrability. A weaker condition for integrability is *Ruelle stability* (Ruelle 1969) meaning that $h(x) \leq \alpha\beta^{n(x)}$ for some positive constants α, β and all $x \in N_{\mathrm{f}}$. As shown later in this section, local stability also plays an important role in simulation algorithms. Local stability is satisfied by many point process models (Geyer 1999, Kendall & Møller 2000). One example, where Ruelle but not local stability is satisfied, is a Lennard-Jones model (Ruelle 1969); this is a pairwise interaction point process (4.30) with $\phi(\xi) = \beta > 0$ constant and $\log\phi(\{\xi,\eta\}) = ar^6 - br^{12}$ for $r = ||\xi - \eta||$, where $a > 0$ and $b > 0$ are parameters.

The role of the Papangelou conditional intensity is similar to that of the local characteristics of a Markov random field when defining local Markov properties. Let $\sim$ be an arbitrary symmetric relation on S, let for the moment $f : N_{\mathrm{f}} \to [0,\infty)$ denote any function, and define λ^* as in (4.33). If f is hereditary and if for any $x \in N_{\mathrm{f}}$ and $\xi \in S \setminus x$, $\lambda^*(x,\xi)$ depends on x only through the *neighbours* $\eta \in x$ to ξ, i.e. those $\eta \in x$ with $\xi \sim \eta$, then f is said to be a *Markov function*. By the *Hammersley-Clifford-Ripley-Kelly theorem* (Ripley & Kelly 1977), f is Markov if and only if f is of the form

$$f(x) = \prod_{y \subseteq x} \phi(y), \quad x \in N_{\mathrm{f}}, \tag{4.34}$$

where ϕ is a so-called *interaction function*, i.e. a function $\phi : N_f \to [0, \infty)$ with the property that $\phi(y) = 1$ if there are two distinct points in y which are not neighbours. If especially f is a density with respect to Poisson(S, μ), we have a *Markov density function* with normalising constant $\phi(\emptyset)$. Then $X \sim f$ is said to be a *Markov point process*. Combining (4.28) and (4.34) we easily obtain a *spatial Markov property*: for Borel sets $A, B \subset S$ so that no point in A is a neighbour to any point in B, X_A and X_B are conditionally independent given X_C where $C = S \setminus (A \cup B)$.

If f is hereditary and for all $x \in N_f$, all $\xi \in S \backslash x$, and some finite $R > 0$ we have that $\lambda^*(x, \xi) = \lambda^*(x \cap b(\xi, R), \xi)$ where $b(\xi, R)$ denotes the closed ball with centre ξ and radius R, then f is said to be of *finite interaction range* R. Then f is obviously Markov with respect to the *finite range neighbour relation* given by $\xi \sim \eta$ if and only if $d(\xi, \eta) \leq R$. So Strauss and multiscale point processes are Markov. For the Norwegian spruce data in Figure 4.2, where a disc $b(\xi, m_\xi)$ specifies the influence zone of a tree located at ξ, it is natural to consider a Markov model with $\sim$ defined by

$$(\xi, m_\xi) \sim (\eta, m_\eta) \Leftrightarrow b(\xi, m_\xi) \cap b(\eta, m_\eta) \neq \emptyset, \tag{4.35}$$

i.e. trees are only allowed to interact when their influence zones overlap. In Section 4.7.5 we consider a Markov model for the spruces with respect to the relation (4.35).

Many other Markov models can be constructed by specifying different kinds of relations and interaction functions, using (4.34) and checking of course for integrability in each case. In fact pairwise interaction point processes are useful models for regularity/inhibition/repulsion but not so much for clustering/attraction. Models for both types of interactions may be constructed by allowing higher order interaction terms, see e.g. Baddeley & van Lieshout (1995), Geyer (1999), and Møller (1999). Moreover, the concept of a Markov function can be extended in different ways, whereby even more flexible models are obtained, see Baddeley & Møller (1989), van Lieshout (2000), and the references therein.

By allowing f to depend on points outside S, it is possible to extend Markov point processes to infinite Gibbs point processes, but questions like existence, uniqueness or not (phase transition behaviour), and stationarity may be hard to answer; we refer the interested reader to Preston (1976) and Georgii (1988) for the mathematical details.

The Papangelou conditional intensity will also play a key role in the sequel concerning statistical inference and simulation procedures for finite point processes.

4.7.2 Pseudo likelihood

Consider a parametric model $f_\theta \propto h_\theta$, $\theta \in \Theta$, for the density of a spatial point process X with respect to $\nu = $ Poisson(S, μ). In general, apart from the Poisson case, the normalising constant

$$\begin{aligned} Z_\theta &= \int h_\theta(x)\nu(\mathrm{d}x) \qquad (4.36) \\ &= \sum_{n=0}^{\infty} \exp(-\mu(S))/n! \int \cdots \int h_\theta(\{x_1,\ldots,x_n\})\mu(\mathrm{d}x_1)\cdots\mu(\mathrm{d}x_n) \end{aligned}$$

cannot be evaluated explicitly. In order to avoid this problem, Besag (1977a) extended the definition of the pseudo likelihood function for Markov random fields (see Besag 1975 and Section 3.4.3) to the Strauss process by an approximation given by an auto-Poisson lattice process (Besag, Milne & Zachary 1982). Based on this derivation a general expression of the pseudo likelihood for point processes is stated in Ripley (1988). The pseudo likelihood for point processes is derived by a direct argument in Jensen & Møller (1991) as follows.

Suppose that each density f_θ is hereditary and Ruelle stable. Let $T \subseteq S$ be an arbitrary Borel set. For $x \in N_\mathrm{f}$, define the *pseudo likelihood* on T by

$$PL_T(\theta;x) = \exp(-\mu(T)) \lim_{i\to\infty} \prod_{j=1}^{m_i} f_\theta(x_{A_{ij}}|x_{S\setminus A_{ij}}),$$

where $\{A_{ij} : j = 1,\ldots,m_i\}$, $i = 1,2,\ldots$, are nested subdivisions of T such that $m_i \to \infty$ and $m_i[\max_{1\le j\le m_i} \mu(A_{ij})]^2 \to 0$ as $i \to \infty$. Further, $f_\theta(x_{A_{ij}}|x_{S\setminus A_{ij}})$ is a conditional density for $X_{A_{ij}}$ given that $X_{S\setminus A_{ij}} = x_{S\setminus A_{ij}}$:

$$f_\theta(x_{A_{ij}}|x_{S\setminus A_{ij}}) = \frac{f_\theta(x_{A_{ij}} \cup x_{S\setminus A_{ij}})}{\int f_\theta(y \cup x_{S\setminus A_{ij}})\mathrm{d}\mu_{A_{ij}}(y)}$$

if the denominator is strictly positive, and $f_\theta(x_{A_{ij}}|x_{S\setminus A_{ij}}) = 0$ otherwise; here $\mu_{A_{ij}}$ denotes the restriction of μ to A_{ij}. By Theorem 2.2 in Jensen & Møller (1991), for μ almost all $x \in N_\mathrm{f}$ the pseudo likelihood on T is well-defined and given by

$$PL_T(\theta;x) = \exp\Big(-\int_T \lambda_\theta^*(x,\xi)\mathrm{d}\mu(\xi)\Big) \prod_{\xi\in x_T} \lambda_\theta^*(x\setminus\{\xi\},\xi) \qquad (4.37)$$

where λ_θ^* is the Papangelou conditional intensity associated to f_θ. Note that the pseudo likelihood function is unaltered whether λ_θ^* is based on f_θ or on the conditional density $f_\theta(x_T|x_{S\setminus T})$. Usually in applications, either $T = S$ or $T = \{\xi \in S : b(\xi,R) \subseteq W\}$ in order to reduce edge effects when $X_W = x$ is observed within a window $W \subseteq S$ and all densities f_θ have finite interaction range R.

The *maximum pseudo likelihood estimate (MPLE)* is found by maximising (4.37). For certain models, consistency and asymptotic normality of the MPLE are established in Jensen & Møller (1991), Jensen & Künsch

(1994), Mase (1995), and Mase (1999). Intuitively, as the pseudo likelihood only depends on the local dependence structure, global information may better be taken into account when using the likelihood function. In fact the asymptotic variance of the MPLE can be much larger than for the MLE (the maximum likelihood estimator), in particular for spatial point processes with high dependence. Note that (4.37) agrees with the likelihood function for a Poisson process when $\lambda_\theta^*(x,\xi)$ depends only on ξ, so for point processes with weak interaction the MPLE and MLE may be expected to be close.

Assume now that the density f_θ belongs to an exponential family,

$$f_\theta(x) = b(x)\exp(\theta \cdot t(x))/Z_\theta, \quad x \in N_{\mathrm{f}},\ \theta \in \Theta, \tag{4.38}$$

where $\Theta = \{\theta \in \mathbb{R}^p : \int b(x)\exp(\theta \cdot t(x))\nu(\mathrm{d}x) < \infty\}$, $\cdot$ is the usual inner product, $b : N_{\mathrm{f}} \to [0,\infty)$ is hereditary, and $t : N_{\mathrm{f}} \to \mathbb{R}^p$. Then

$$\lambda_\theta^*(x,\xi) = b(x,\xi)\exp(\theta \cdot t(x,\xi)) \tag{4.39}$$

where $b(x,\xi) = b(x \cup \xi)/b(x)$ and $t(x,\xi) = t(x \cup \xi) - t(x)$. Proposition 2.3 in Jensen & Møller (1991) states that $PL_T(\theta;x)$ is log concave, and gives a condition for strict log concavity. Hence, if the MPLE exists and belongs to the interior of Θ, it is the solution to the pseudo likelihood equation $(\mathrm{d}/\mathrm{d}\theta)\log PL_T(\theta;x) = 0$, which is equivalent to

$$\int_T b(x,\xi)t(x,\xi)\exp(\theta \cdot t(x,\xi))\mathrm{d}\mu(\xi) = \sum_{\xi \in x_T} t(x \setminus \{\xi\},\xi). \tag{4.40}$$

Maximum pseudo likelihood estimation based on (4.37) is computationally equivalent to maximum likelihood estimation in an inhomogeneous Poisson process. In practice the integrals in (4.37) and (4.40) may be approximated by numerical methods. As noticed in Berman & Turner (1992) and Baddeley & Turner (2000), standard software such as Splus for fitting generalised linear models can be used to provide an approximate MPLE as follows.

Consider again the case (4.39), partition T into cells C_i, and let $c_i \in C_i$ denote a given "centre point". Let u_j, $j = 1,\dots,m$, denote a list of these centre points and the points in x_T. Then the integral in (4.37) is approximated by

$$\int_T \lambda_\theta^*(x,\xi)\mathrm{d}\mu(\xi) \approx \sum_{j=1}^m \lambda_\theta^*(x \setminus u_j, u_j)w_j, \tag{4.41}$$

where $w_j = \mu(C_i)/(1 + x(C_i))$ if $u_j \in C_i$. Note that the approximation (4.41) involves a "discontinuity error", since for $u_j \in x$, $\lambda_\theta^*(x \setminus u_j, u_j)$ is in general not equal to the limit of $\lambda_\theta^*(x,\xi)$ as $\xi \to u_j$, cf. the discussion in

Baddeley & Turner (2000). The advantage of including x_T in the sum in (4.41) is that we obtain

$$\log PL_T(\theta; x) \approx \sum_{j=1}^{m} (y_j \log \lambda_j^* - \lambda_j^*) w_j, \tag{4.42}$$

where $y_j = 1[u_j \in x]/w_j$ and $\lambda_j^* = \lambda_\theta^*(x \setminus u_j, u_j)$. The right side of (4.42) is formally equivalent to the log likelihood of independent Poisson variables y_j with means λ_j^* taken with weights w_j. If $b(x, u_j) > 0$, $j = 1, \ldots, m$, then (4.42) can easily be maximised using standard software for generalised linear models (taking $\log b(x, u_j)$ as an offset term). Moreover, if $b(\cdot, \cdot) = 1$, then we have a log linear model

$$\log \lambda_j^* = \theta \cdot t(x \setminus u_j, u_j).$$

4.7.3 Example 2: Norwegian spruces (continued)

For the point pattern of spruce locations we consider a multiscale process (4.32) with $k = 5$ and $R_i = 1.1 \times i$, $0 < \gamma_i \leq 1$, $i = 1, \ldots, 4$. The minimal interaction range is thus less or equal to 4.4, a value suggested by the estimated summary statistics in Figure 4.3. We use the approximation (4.41) with $T = [0, 56] \times [0, 38]$ partitioned into 56×38 quadratic cells C_i, $i = (k, l) \in \{0, \ldots, 55\} \times \{0, \ldots, 37\}$, each of unit area and with centre points $c_{(k,l)} = ((k + 0.5)/56, (l + 0.5)/38)$. We have a log linear model with $\theta = (\log \beta, \log \gamma_1, \ldots, \log \gamma_4) \in \mathbb{R} \times (-\infty, 0]^4$ and $t(x, u_j) = (1, s[j, 1], \ldots, s[j, 4])$, where $s[j, i]$ denotes the number of points $\xi \in x \setminus u_j$ with $R_{i-1} < \|\xi - u_j\| \leq R_i$, $i = 1, \ldots, 4$. Using the Splus routine `glm()` with the call

```
glm(y~s[,1]+...+s[,4],family=poisson(link=log),weights=w),
```

and with y, w, and s constructed as above, the estimates -0.84, -3.35, -1.38, -0.62, -0.15 for $\log \beta$, $\log \gamma_1$, ..., $\log \gamma_4$ are obtained (for a comparison with maximum likelihood estimates, see Section 4.7.5).

A biologically more interesting model is obtained by treating the spruce data as a marked point pattern where the stem diameters are used in the modelling. This approach is considered in Goulard et al. (1996) who discuss pseudo likelihood inference for marked point processes with the spruce data as one of the examples. We further discuss a marked point process approach in Section 4.7.5.

4.7.4 Likelihood inference

Consider again a parametric model of densities $f_\theta \propto h_\theta$, $\theta \in \Theta$, with respect to Poisson(S, μ), and where a closed expression for the normalising constant Z_θ given by (4.36) is not avaliable. In this section we discuss how to find the *maximum likelihood estimator (MLE)* $\hat{\theta}$ and the *likelihood ratio statistic* for hypothesis testing using MCMC methods. For simplicity we assume that the support $\{x : h_\theta(x) > 0\}$ does not depend on $\theta \in \Theta$. Furthermore, $\mathbb{E}_\theta$ denotes expectation with respect to $X \sim f_\theta$.

Assume that a realization $X = x$ is observed. In the exponential family case (4.38), $\hat{\theta}$ is the solution to the likelihood equation $\mathbb{E}_\theta t(X) = t(x)$. This suggest to approximate $\mathbb{E}_\theta t(X)$ by Monte Carlo methods, e.g. combined with Newton-Raphson (Penttinen 1984) or the EM-algorithm or stochastic approximation/gradient methods. Geyer (1999) advocates the use of other methods based on *importance sampling* as described below; see Geyer & Thompson (1992), Geyer & Møller (1994), Gu & Zhu (2001), and the references therein; see also Sections 1.5.4 and 3.4. Suppose that $\psi \in \Theta$ is an initial guess of $\hat{\theta}$, e.g. the MPLE. Then

$$Z_\theta / Z_\psi = \mathbb{E}_\psi \big[h_\theta(X) / h_\psi(X) \big] \tag{4.43}$$

can be estimated by a sample $X_1, X_2, \ldots$ from an irreducible Markov chain with invariant density f_ψ, see Section 4.7.7. Hence the logarithm of the likelihood ratio

$$f_\theta(x)/f_\psi(x) = \big(h_\theta(x)/h_\psi(x)\big) / \big(Z_\theta/Z_\psi\big)$$

is approximated by

$$l_n(\theta) = \log\left[\frac{h_\theta(x)}{h_\psi(x)}\right] - \log\left[\frac{1}{n}\sum_{i=1}^{n} \frac{h_\theta(X_i)}{h_\psi(X_i)}\right]. \tag{4.44}$$

For fixed ψ we can consider (4.44) as an approximation of the log *likelihood function* from which we may obtain an *approximate MLE* $\hat{\theta}_n$. Defining the importance weights

$$w_{\theta,\psi,n}(x) = \frac{h_\theta(x)/h_\psi(x)}{\sum_{i=1}^{n} h_\theta(X_i)/h_\psi(X_i)}$$

and for any function $k : N_{\mathrm{f}} \to \mathbb{R}^p$,

$$\mathbb{E}_{\theta,\psi,n} k(X) = \sum_{i=1}^{n} k(X_i) w_{\theta,\psi,n}(X_i),$$

we obtain approximate score functions, etc. by replacing exact expectations by Monte Carlo expectations. For example, in the exponential family case (4.38) the *score function* is approximated by

$$\nabla l_n(\theta) = t(x) - \mathbb{E}_{\theta,\psi,n} t(X), \tag{4.45}$$

and the *Fisher information* by

$$-\nabla^2 l_n(\theta) = \mathbb{V}\mathrm{ar}_{\theta,\psi,n} t(X), \tag{4.46}$$

and $l_n(\theta)$ is concave so that Newton-Raphson is feasible. Note that $\hat{\theta}_n$ is a function of both x and $X_1, \ldots, X_n$. Asymptotic normality of the Monte Carlo error $\sqrt{n}(\hat{\theta}_n - \hat{\theta})$ as $n \to \infty$ is established in Geyer (1994).

The approximations (4.44)–(4.46) are only useful for θ sufficiently close to ψ. When ψ is not close to $\hat{\theta}$, Geyer & Thompson (1992) propose to use an iterative procedure with a "trust region", but one should be particular careful if the likelihood function is multi-modal.

A natural requirement is that $l_n(\theta)$ has finite variance. This is the case if the chain is geometrically ergodic (Section 1.4.3) and $\mathbb{E}_\psi |h_\theta(X)/h_\psi(X)|^{2+\epsilon} < \infty$ for some $\epsilon > 0$ (Theorem 1 in Chan & Geyer (1994)); or if just $\mathbb{E}_\psi(h_\theta(X)/\ h_\psi(X))^2 < \infty$ provided the chain is reversible (Corollary 3 in Roberts & Rosenthal (1997)). For example, for the Strauss process (4.31), if $\theta = (\beta, \gamma, R)$ and $\psi = (\beta', \gamma', R')$ with $\beta, \beta', R, R' > 0$ and $\gamma, \gamma' \in [0,1]$, then $\mathbb{E}_\psi |h_\theta(X)/h_\psi(X)|^2 < \infty$ if and only if $\gamma \le \sqrt{\gamma'}$.

In order to estimate Z_θ/Z_ψ and hence the log likelihood ratios when θ and ψ are far apart umbrella sampling and the method of reverse logistic regression have been proposed, see Geyer (1991) and Geyer (1999). It is however our experience that these methods are often numerically unstable due to large variances for ratios of unnormalised densities. We turn therefore now to another technique called *path sampling*. The advantages of using this approach over the importance sampling approach is discussed in Gelman & Meng (1998).

Briefly, path sampling works as follows (see also Section 1.7.2). Suppose that $\Theta \subseteq \mathbb{R}^p$, $\theta \to \log h_\theta(X)$ is differentiable for $\theta \in \Theta$, and $(\mathrm{d}/\mathrm{d}\theta) h_\theta(X)$ is locally dominated integrable along a continuous differentiable path $\theta(s) \in \Theta$, $0 \le s \le 1$ where $\psi = \theta(0)$ and $\theta = \theta(1)$. Letting

$$V_\theta(X) = (\mathrm{d}/\mathrm{d}\theta) \log h_\theta(X) \quad \text{and} \quad \theta'(s) = \mathrm{d}\theta(s)/\mathrm{d}s,$$

the identity

$$\log\left(Z_\theta/Z_\psi\right) = \int_0^1 \mathbb{E}_{\theta(s)} V_{\theta(s)}(X) \theta'(s)^{\mathsf{T}} \mathrm{d}s \tag{4.47}$$

is straightforwardly derived. Note that for many exponential family models (4.38), Monte Carlo estimation is more stable for $\mathbb{E}_{\theta(s)} V_{\theta(s)}(X) = \mathbb{E}_{\theta(s)} t(X)$ in (4.47) than for $\mathbb{E}_\psi[h_\theta(X)/h_\psi(X)] = \mathbb{E}_\psi \exp((\theta - \psi) \cdot t(X))$ in (4.43). The right hand side in (4.47) can be approximated by a Riemann sum using a discrete grid of points $\theta(s_i)$, $i = 1, \ldots, m$, generating independent Markov chains X_t^i, $i = 1, \ldots, m$, with invariant densities $f_{\theta(s_i)}$, and estimating $\mathbb{E}_{\theta(s_i)} V_{\theta(s_i)}(X)$ by Monte Carlo; Berthelsen & Møller (2001b) combine this with independent runs of the chains, starting with a perfect simulation (see Section 4.7.9) for each chain. Alternatively, a Markov chain (X_t, S_t) defined on $E \times [0,1]$ may be used; see Gelman & Meng (1998) for details.

Often in applications we can choose ψ or θ so that Z_ψ or Z_θ is known. For example, for the Strauss process (4.31), when (β, R) is fixed and $\Theta = \{\gamma \in (0,1]\}$, we may choose $\theta = 1$ so that Z_θ is the normalising constant of a Poisson process. Then by (4.47), for $0 < \psi < 1$,

$$\log\left(Z_1/Z_\psi\right) = \int_\psi^1 \mathbb{E}_\gamma \sum_{\{\xi,\eta\}\subseteq x} 1[d(\xi,\eta) \le R]/\gamma \,\mathrm{d}\gamma. \tag{4.48}$$

Thereby an estimate of $\log Z_\psi$ is obtained. Repeating this for different values of (β, R), the entire likelihood surface, the likelihood ratio statistic for a specified hypothesis (e.g. that $\gamma = 1$), etc., can be approximated. For details, see Berthelsen & Møller (2001b) who also determine the distribution of the approximate likelihood ratio statistic by making further perfect and independent simulations.

Similar methods apply for *missing data situations.* Suppose that only $X_W = x$ is observed within a window $W \subseteq S$. Let $V = S\setminus W$, $Y = X_W$, and $Z = X_V$ (which is unobserved). Recall that if $X \sim \nu = \text{Poisson}(S,\mu)$, then $Y \sim \nu_W = \text{Poisson}(W,\mu_W)$ and $Z \sim \nu_V = \text{Poisson}(V,\mu_V)$ are independent, where μ_A denotes the restriction of μ to A. So if $X \sim f_\theta$, then Y has density

$$f_{\theta,W}(x) = \int f_\theta(x \cup z)\nu_V(\mathrm{d}z)$$

with respect to ν_W. Note that $Z_\theta(x) = \int h_\theta(x\cup z)\nu_V(\mathrm{d}z)$ is the normalising constant of the conditional density $f_{\theta,V}(z|x) \propto h_\theta(x \cup z)$ with respect to ν_V. Consequently, the logarithm of the likelihood ratio

$$f_{\theta,W}(x)/f_{\psi,W}(x) = \left[Z_\theta(x)/Z_\psi(x)\right]/\left[Z_\theta/Z_\psi\right]$$

can be approximated by

$$l_{n,W}(\theta) = \log\left[\frac{1}{n}\sum_{i=1}^n \frac{h_\theta(x \cup Z_i)}{h_\psi(x \cup Z_i)}\right] - \log\left[\frac{1}{n}\sum_{i=1}^n \frac{h_\theta(X_i)}{h_\psi(X_i)}\right] \tag{4.49}$$

where $Z_1, Z_2, \ldots$ is a sample from an irreducible Markov chain with invariant density $f_{\theta,V}(z|x)$, and $X_1, X_2, \ldots$ is a chain as in (4.44). Alternatively, path sampling can be used for estimating each of the terms $\log\left[Z_\theta/Z_\psi\right]$ and $\log\left[Z_\theta(x)/Z_\psi(x)\right]$.

4.7.5 Example 2: Norwegian spruces (continued)

For the spruce data we consider two different models: first a multiscale model as in Section 4.7.3 where we ignore the stem diameters and, second, a biologically more realistic model where overlap of the influence zones (see Section 4.2.2 and Figure 4.2) of the trees is penalised.

For the multiscale process the likelihood is maximised using Newton-Raphson with the score function and Fisher information approximated by (4.45) and (4.46), respectively. We apply a trust region procedure where ψ and θ initially are taken equal to the maximum likelihood estimate under the Poisson model (i.e. with $\log \gamma_i = 0$, $i = 1, \ldots, 4$). When the output $\tilde{\theta}$ of a Newton-Raphson iteration falls outside the trust region $\prod_{i=1}^{5}[\psi_i - 0.05, \psi_i + 0.05]$ new approximations (4.45) and (4.46) are calculated with the previous ψ value replaced by $\tilde{\theta}$. The Newton-Raphson procedure converges to the estimate $\hat{\theta} = (-0.38, -3.69, -1.49, -0.71, -0.30)$ for which the repulsion is stronger than for the pseudo likelihood estimate obtained in Section 4.7.3. We also maximise the likelihood under the null hypothesis $\gamma_1 = \gamma_2 = \gamma_3 = \gamma_4 = \gamma$, that is for the Strauss process with $R = 4.4$, and obtain the estimate $(-1.27, -0.44)$ for $(\log \beta, \log \gamma)$. The left plot in Figure 4.8 shows the interaction functions corresponding to $\hat{\theta}$ and the estimate under the null hypothesis.

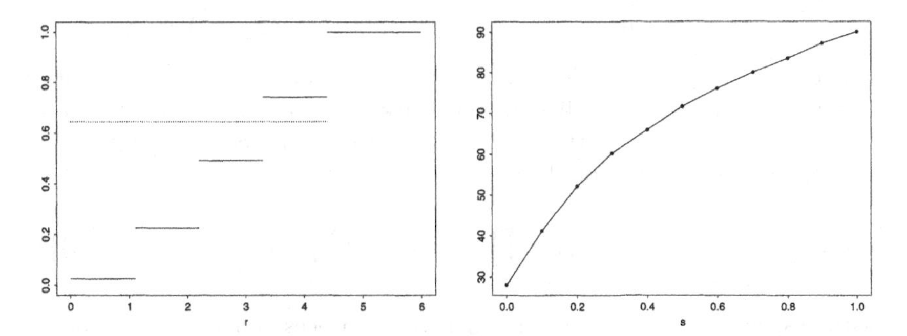

FIGURE 4.8. Left: estimated multiscale interaction function $\phi(\{\xi, \eta\})$ (see (4.32)) plotted as a function of distance $r = \|\xi - \eta\|$; solid line is for $\hat{\theta}$ and dotted is for the estimate under the null hypothesis (the Strauss model). Right: Monte Carlo estimates of $E_{\theta(k/10)} V_{\theta(k/10)}(X)\theta'(k/10)^{\mathsf{T}}$, $k = 0, \ldots, 10$ and curve corresponding to trapezoidal approximation.

Using path sampling we finally compute the log likelihood ratio statistic for the null hypothesis. Letting $\theta_0 = (-1.27, -0.44, -0.44, -0.44, -0.44)$ and $\theta_1 = \hat{\theta}$, we use the path $\theta(s) = \theta_0 + (\theta_1 - \theta_0)s$. The integral (4.47) is approximated as follows: for each of 11 quadrature points $\theta(k/10)$, $k = 0, \ldots, 10$, the integrand values $\mathrm{E}_{\theta(k/10)} V_{\theta(k/10)}(X)\theta'(k/10)^{\mathsf{T}}$ are replaced by Monte Carlo estimates and the integral with respect to s is finally approximated using the trapezoidal rule. The right plot in Figure 4.8 shows the Monte Carlo estimates of $\mathrm{E}_{\theta(k/10)} V_{\theta(k/10)}(X)\theta'(k/10)^{\mathsf{T}}$ together with the trapezoidal approximation. The log ratio $\log Z(\theta_1)/Z(\theta_0)$ is equal to 68 and the value of -2 times the log likelihood ratio statistic is 44 which is highly significant according to standard asymptotic results for the likelihood ratio statistic. Alternatively one may consider a parametric bootstrap where the observed log likelihood ratio statistic is compared with the distribution of the log likelihood ratio statistic under the fitted Strauss model. Specifically we compute -2 times the log likelihood ratio statistic from 99 simulations under the fitted Strauss model (proceeding exactly as for the observed data), obtaining values between 0.06 and 15.66, so the bootstrap also provides strong evidence against the Strauss model.

Neither the multiscale nor the Strauss model are very satisfactory from a biological point of view. A more realistic model can be obtained if the influence zones are included in the modelling so that low probability is assigned to point patterns with large area of overlaps between the influence zones. Let m_{lo} and m_{up} denote the minimal and maximal observed radii for the influence zones. For marked points (x_1, m_1) and (x_2, m_2) in $S = [0,56] \times [0,38] \times [m_{\mathrm{lo}}, m_{\mathrm{up}}]$ we define interaction functions ϕ_1 and ϕ_2 by

$$\phi_1((x_i, m_i)) = \beta_k \quad \text{if } k(m_{\mathrm{up}} - m_{\mathrm{lo}})/6 < m_i - m_{\mathrm{lo}} \leq (k+1)(m_{\mathrm{up}} - m_{\mathrm{lo}})/6$$

where $\beta_k > 0$ and $k = 0, \ldots, 5$, and

$$\log \phi_2(\{(x_1, m_1), (x_2, m_2)\}) = |b(x_1, m_1) \cap b(x_2, m_2)| \log \gamma$$

where $0 < \gamma \leq 1$. The function ϕ_1 allows modelling of the intensities of points with different values of the marks and ϕ_2 models the degree of repulsion in the point pattern. The marked point process is finally given by the pairwise interaction density

$$f(\{(x_1, m_1), \ldots, (x_n, m_n)\}) \propto \prod_{i=1}^{n} \phi_1((x_i, m_i)) \prod_{i<j} \phi_2(\{(x_i, m_i), (x_j, m_j)\})$$

with respect to the standard Poisson point process on $[0,56] \times [0,38] \times [m_{\mathrm{lo}}, m_{\mathrm{up}}]$. Proceeding as for the multiscale process using Newton-Raphson and path sampling we compute maximum likelihood estimates and the likelihood ratio statistic for the null hypothesis $\beta_0 = \beta_1 = \cdots = \beta_5 = \beta$. The maximum likelihood estimates are $(-1.23, -0.34, 0.53, -0.40, -0.80, -0.67, -1.10)$ for $(\log \beta_0, \ldots, \log \beta_5, \log \gamma)$ in the full model and $(-0.33, -1.07)$ for $(\log \beta, \log \gamma)$ under the reduced model. The log likelihood ratio statistic is -25 which is highly significant according to both standard asymptotics and a parametric bootstrap.

We conclude by giving some computational details. The samples used in the Newton-Raphson optimisation for Monte Carlo estimation of the score function and Fisher information were of length 5000 and obtained by subsampling each 200th state of a Metropolis-Hastings chain generated as described in Section 4.7.7. The Monte Carlo estimates of the integrands $\mathbb{E}_{\theta(k/10)} V_{\theta(k/10)}(X) \theta'(k/10)^{\mathsf{T}}$ in the path sampling procedure were computed from samples of length 1000 also obtained by subsampling each 200th state of a Metropolis-Hastings chain. The influence of the sample lengths and subsampling intervals on the Monte Carlo error of the various Monte Carlo estimates is a subject that deserves further study. The simulations for the parametric bootstrap were obtained by subsampling each 1000th state of a Metropolis-Hastings chain to obtain approximately uncorrelated simulations. A more appropriate approach would be to generate independent samples using perfect simulation, see Section 4.7.9.

4.7.6 *Bayesian inference*

Below we shortly comment on some of the rather few Bayesian contributions for Markov point processes which so far have been published.

Suppose we are extending the situation considered at the beginning of the previous section to a Bayesian setting with a prior on θ. The posterior distribution for $\theta|X = x$ is complicated by the fact that the normalising constant Z_θ in the likelihood term is usually unknown. Heikkinen & Penttinen (1999) suggest a Bayesian smoothing technique for estimation in pairwise interaction processes, where the likelihood function is approximated by the multiscale point process (4.32) having a large number of fixed change points $R_1, \ldots, R_{k-1}$. For convenience, they condition on the observed number $n(x)$ of points. A Gaussian Markov chain prior for $\theta = (\log \gamma_1, \ldots, \log \gamma_{k-1})$ is chosen so that large differences $|\log \gamma_i - \log \gamma_{i-1}|$ are penalised. As the full posterior analysis is considered to be too demanding, they concentrate on finding the posterior mode, using ideas from MCMC MLE as given in Penttinen (1984) and Geyer & Thompson (1992). Berthelsen & Møller (2001a) consider a similar situation, without conditioning on $n(x)$ but imposing a prior on k and $R_1, \ldots, R_{k-1}$, and finding the normalising constant of the likelihood term by path sampling so that a full Bayesian analysis is possible.

Lund, Penttinen & Rudemo (1999) consider a situation where an unobserved point process X is degraded by independent thinning, random displacement, a simple censoring mechanism, and independent superpositioning with a Poisson process of "ghost points"; this is related to aerial photographs of trees disturbed by the image analysis process, cf. Lund & Rudemo (2000). A known pairwise interaction prior on X is imposed in Lund et al. (1999), so its normalising constant is unimportant when dealing with the posterior distribution for X. Perfect simulation for this posterior is discussed in Lund & Thönnes (2000) and Møller (2001).

Finally, we mention in passing that Bayesian cluster models, using a locally stable prior density for the mother points, are studied in Baddeley & van Lieshout (1993), van Lieshout (2000), and Loizeaux & McKeague (2001). Here the parameters for the prior density for the mother points are assumed to be known, and local stability of the posterior density may be established. In particular, Loizeaux & McKeague (2001) discuss perfect simulation for the posterior distribution, and applies this to data on cases of leukaemia.

4.7.7 *Metropolis-Hastings algorithms*

In this and the following sections we concentrate on how to make simulations from a finite spatial point process X with a density f with respect to Poisson(S, μ) where $0 < \mu(S) < \infty$. Conditional simulation given that the number of points $n(X) = n$ is fixed may be done by any standard algorithm for updating n components $(X_1, \ldots, X_n)$ with density $f_n(x_1, \ldots, x_n) \propto$

$f(\{x_1, \ldots, x_n\})$ with respect to the product measure $\mu \times \cdots \times \mu$ (n times), e.g. by the classical *Metropolis algorithm* (Metropolis, Rosenbluth, Rosenbluth, Teller & Teller 1953) or by *Gibbs sampling* (Ripley 1979); see the survey in Møller (1999). Incidentally the purpose in Metropolis et al. (1953) was to simulate a hard core point process, and also the Gibbs sampler, which is now widely used in statistics, was first introduced in spatial statistics (and in statistical physics) in connection to random fields and spatial point processes.

We follow Geyer & Møller (1994) and define a *Metropolis-Hastings* chain $X_0, X_1, \ldots$ as follows. For technical reasons, assume that f is hereditary, cf. (4.29). Define the state space E of the chain as the set of finite point configurations which are feasible with respect to f, i.e. $E = \{x \in N_\mathrm{f} : f(x) > 0\}$. Let $\bar{\mu} = \mu/\mu(S)$ denote the normalisation of the measure μ. Now, for $X_{i-1} = x \in E$, with probability 1/2 we propose to add a point $\xi \sim \bar{\mu}$ to x, and else we generate a uniformly selected point $\eta \in x$ and propose to delete η from x (if $x = \emptyset$ we set $\eta = \emptyset$). In the former case we return $X_i = x \cup \xi$ with probability $\min\{1, r(x,\xi)\}$ where $r(x,\xi) = \lambda^*(x,\xi)\mu(S)/(n(x)+1)$, and we retain $X_i = x$ otherwise. In the latter case we return $X_i = x \setminus \eta$ with probability $\min\{1, 1/r(x \setminus \eta, \eta)\}$ (setting this to 1 if $x = \emptyset$), and we retain $X_i = x$ otherwise.

The algorithm provides a simple example of Peter Green's reversible jump MCMC algorithm (Green 1995, Waagepetersen & Sorensen 2001). Its theoretical properties are studied in Geyer & Møller (1994), Geyer (1999), and Møller (1999). The chain is straightforwardly seen to be reversible with respect to f. Since f is hereditary, the state $\emptyset$ can be reached with probability 1 within a finite number of steps from any other state $x \in E$. It thereby follows that the chain is irreducible, and that f specifies the unique invariant distribution. As it can stay in $\emptyset$ with a positive probability, it is aperiodic. Assuming local stability, geometrical ergodicity of the chain can be established, and it becomes uniformly ergodic if and only if $f(x) = 0$ whenever $n(x)$ is sufficiently large, cf. the above-mentioned references.

The algorithm can obviously be modified by using other kinds of proposals for the addition or deletion of a point, and by incorporating the possibility of making a "fixed dimension move" as mentioned at the beginning of this section, cf. Geyer & Møller (1994). Such modifications may improve the mixing properties of the chain, but one should keep in mind that extra programming will be needed and the CPU time for each transition will usually be increased. The algorithm may also be combined with auxiliary variable techniques. For example, it is combined with *simulated tempering* in Mase, Møller, Stoyan, Waagepetersen & Döge (2001) in order to make simulations of hard core point processes with a high density of points feasible.

4.7.8 Simulations based on spatial birth-death processes

Let the situation be as at the beginning of the previous section. Preston (1977) notices that under suitable conditions, (approximate) realizations of $X \sim f$ may be obtained by running a *spatial birth-death process* $Y = \{Y_t : t \geq 0\}$ for a sufficient long time (briefly, Y is a continuous time Markov chain where each transition consists in either adding a new point or deleting an existing point). In this section we consider a *coupling construction* for the simplest case, which becomes useful for making simulations; this construction is also used in Section 4.7.9 for making perfect simulations.

Assume again that local stability is satisfied and let $K \geq \lambda^*$ denote an upper bound on the Papangelou conditional intensity. We start by describing how a spatial birth-death process $D = \{D_t : t \geq 0\}$ with equilibrium distribution Poisson$(S, K\mu)$ can easily be generated. This is next used to generate the abovementioned process Y by a thinning procedure so that $D_t \supseteq Y_t$ for all $t \geq 0$, provided $D_0 \supseteq Y_0$; we say that D *dominates* Y.

Generation of D: Given an initial state $D_0 \in N_f$, it is straightforward to generate successive transition times and new states for D as follows. Suppose that we have generated D_s for $0 \leq s \leq t$, and $D_t = x = \{x_1, \ldots, x_n\}$. Let $\tau_0, \tau_1, \ldots, \tau_n$ be independent and exponentially distributed with means $1/(K\mu(S)), 1, \ldots, 1$, respectively. Then the waiting time until the next transition in D is given by $\tau = \min\{\tau_0, \tau_1, \ldots, \tau_n\} \sim \text{Exp}(K\mu(S) + n)$. If $\tau = \tau_0$, then we have a birth: generate a point $\xi \sim \bar{\mu} = \mu/\mu(S)$ and set $D_{t+\tau} = D_t \cup \xi$. If instead $\tau = \tau_i$ with $1 \leq i \leq n$, then we have a death: set $D_{t+\tau} = D_t \setminus x_i$.

It can be shown that no matter which initial state $D_0 \in N_f$ is used, D converges towards its equilibrium distribution given by $\nu = \text{Poisson}(S, K\mu)$: for any $F \in \mathcal{N}_f$, $P(D_t \in F | D_0) \to \nu(F)$ as $t \to \infty$. Moreover, D regenerates each time $D_t = \emptyset$ and $D_{t-} \neq \emptyset$ (where $D_{t-} = \lim_{s \uparrow t} D_s$), and with probability 1, this happens infinitely often.

Generation of Y: Given initial states $Y_0 \in E$ and $D_0 \in N_f$ with $Y_0 \subseteq D_0$, and letting the transition times of Y be included in the transition times of D, it is straightforward to generate successive transition times and new states for Y by thinning from D as follows. Suppose that we have generated $Y_s \subseteq D_s$ for $0 \leq s \leq t$, where every $Y_s \in E$, and $D_t = x = \{x_1, \ldots, x_n\}$ and $Y_t = y \subseteq x$. Using a notation as above, in the case of a birth $D_{t+\tau} = D_t \cup \xi$, we let $Y_{t+\tau} = Y_t \cup \xi$ with probability $\lambda^*(y, \xi)/K$, and retain $Y_{t+\tau} = Y_t$ otherwise. In the case of a death $D_{t+\tau} = D_t \setminus x_i$, we set $Y_{t+\tau} = Y_t \setminus x_i$; so Y is unchanged at time $t + \tau$ if $x_i \notin Y_t$.

It can be shown that for all initial states $(D_0, Y_0) \in N_f \times E$ with $D_0 \supseteq Y_0$, Y converges towards its equilibrium density f as $t \to \infty$. Further, Y regenerates each time $Y_t = \emptyset$ and $Y_{t-} \neq \emptyset$, and with probability 1, this happens infinitely often. Hence by the renewal theorem, for any measurable function $k : E \to [0, \infty)$,

$$\frac{1}{t} \int_0^t k(Y_s) \mathrm{d}s \to \mathbb{E}k(X) \quad \text{as } t \to \infty \tag{4.50}$$

almost surely. Finally, D and Y are each reversible, but (D, Y) is in general not reversible. See Preston (1977) and Møller (1989) for further details, and Berthelsen & Møller (2002) for extensions to more general cases of spatial birth-death processes and other kinds of spatial jump processes.

If Y is generated on a finite time interval $[0, t]$, (4.50) may be used for estimating expectations. However, Metropolis-Hastings simulations as described in Section 4.7.7 seem more popular in practice, possibly due to their simplicity. Spatial birth-death processes have advantages for perfect simulation as will be demonstrated in the following section.

4.7.9 Perfect simulation

Since the seminal paper by Propp & Wilson (1996), *perfect or exact simulation* has been an intensive area of research. The term "perfect simulation" rather than "exact simulation" has been introduced in Kendall (1998) to emphasise that the output of the algorithms are only exact up to deficiencies in the pseudo random number generator applied in the computer implementation of the algorithm, and since very long runs may be omitted due to time constraints, which possibly is causing a bias in the output, cf. the discussion in Kendall & Møller (2000).

Perfect simulation techniques seem particular applicable for many spatial point process models, see e.g. Kendall (1998), Häggström, van Lieshout & Møller (1999), Thönnes (1999), Kendall & Møller (2000), Berthelsen & Møller (2001a); see also the surveys Møller (2001), Berthelsen & Møller (2001b), and the references therein. In this section we follow Kendall (1998) and Kendall & Møller (2000) and show how the coupling construction of the spatial birth-death processes Y and D introduced in Section 4.7.8 can be used for making perfect simulations from a locally stable density f.

Recall that $\{D_t : t \geq 0\}$ is reversible with invariant distribution $\nu = \text{Poisson}(S, K\mu)$. Hence we can easily start in equilibrium $D_0 \sim \nu$, and extend the process backwards in time to obtain $\{D_t : t \leq 0\}$, using the same procedure as for forwards simulations of D. Let $\ldots, T_{-2} < T_{-1} < T_1 < T_2 < \ldots$ denote the times where $D_t = \emptyset$ and $D_{t-} \neq \emptyset$, such that $T_{-1} \leq 0 < T_1$. As D regenerates at these time instances, the cycles of D

$$\ldots, \{D_t : T_{-2} \leq t < T_{-1}\}, \{D_t : T_{-1} \leq t < T_1\}, \{D_t : T_1 \leq t < T_2\}, \ldots$$

are i.i.d. Imagine that we generate Y within each cycle of D, setting first $Y_{T_i} = \emptyset$ for $i \in \mathbb{Z} \setminus \{0\}$ as D dominates Y, and then using the forwards thinning procedure described in Section 4.7.8. Then $\{(D_t, Y_t) : -\infty < t <$

$\infty\}$ is a continuous time stationary process. Consequently, for any fixed time t, we have that $Y_t \sim f$.

This means that a perfect simulation $Y_0 \sim f$ can be obtained by first simulating $\{D_t : 0 \geq t \geq T_{-1}\}$ backwards in time, starting in equilibrium at time 0, and then generate $\{Y_t : T_{-1} \leq t \leq 0\}$ forwards in time by the thinning procedure. For this we actually only need to generate the jump chain of $\{D_t : 0 \geq t \geq T_{-1}\}$, i.e. the states D_t where a backwards transition occurs, since this contains the jump chain of $\{Y_t : T_{-1} \leq t \leq 0\}$. However, T_{-1} can be infeasible large, cf. Berthelsen & Møller (2001b), so alternative algorithms as described below are used in practice.

One possibility is to use *upper and lower processes* defined as follows. Let again $D_0 \sim \nu$, denote $Z_{-1}, Z_{-2}, \ldots$ the jump chain of $\{D_t : t < 0\}$ when considered backwards in time, and denote $\ldots, W_{-2}, W_{-1}$ the states of $\{Y_t : t < 0\}$ at the times where the jumps of $\{D_t : t < 0\}$ occur when considered forwards in time. Note that $D_0 = Z_{-1}$, $Y_0 = W_{-1}$, and the jump chain of $\{Y_t : t < 0\}$ agrees with the jumps of $\ldots, W_{-2}, W_{-1}$. For $n = 1, 2, 3\ldots$, we define below the upper process $U^n = \{U_t^n : t = -n, \ldots, -1\}$ and the lower process $L^n = \{L_t^n : t = -n, \ldots, -1\}$ so that

$$L_t^n = U_t^n \;\Rightarrow\; L_s^n = U_s^n \quad \text{for } s = t, \ldots, -1, \tag{4.51}$$

and

$$L_t^n \subseteq W_t \subseteq U_t^n \subseteq Z_t. \tag{4.52}$$

The *coalescence property* (4.51) and the *sandwiching property* (4.52) imply that if $L_t^n = U_t^n$ for some $-n \leq t \leq 0$, then by induction $L_s^n = W_s = U_s^n$ for $s = t, \ldots, -1$, and so $U_{-1}^n = W_{-1} = Y_0 \sim f$. Hence,

$$L_t^n = U_t^n \text{ for some } -n \leq t \leq 0 \;\Rightarrow\; U_{-1}^n \sim f.$$

We now consider the coupling construction for Z and W, and thereby realize how to extend this to a coupling construction for upper and lower processes satisfying (4.51) and (4.52). Let $R_{-1}, R_{-2}, \ldots$ be independent and uniformly distributed on $[0,1]$, and independent of $Z_{-1}, Z_{-2}, \ldots$. For each n, since we only know that $\emptyset \subseteq W_{-n} \subseteq Z_{-n}$, we set first $U_{-n}^n = Z_{-n}$ and $L_{-n}^n = \emptyset$. Then we iterate as follows for $t = -n+1, \ldots, -1$: If a death happens so that $Z_t = Z_{t-1} \setminus \eta$, then $W_t = W_{t-1} \setminus \eta$, and so we set $U_t^n = U_{t-1}^n \setminus \eta$ and $L_t^n = L_{t-1}^n \setminus \eta$. If instead a birth $Z_t = Z_{t-1} \cup \xi$ happens, then $W_t = W_{t-1} \cup \xi$ if $R_t < \lambda^*(W_{t-1}, \xi)/K$, while $W_t = W_{t-1}$ is unchanged otherwise; so we set

$$U_t^n = U_{t-1}^n \cup \xi \quad \text{if } R_t < \max\{\lambda^*(x, \xi)/K : L_{t-1}^n \subseteq x \subseteq U_{t-1}^n\} \tag{4.53}$$

and $U_t^n = U_{t-1}^n$ otherwise, and set

$$L_t^n = L_{t-1}^n \cup \xi \quad \text{if } R_t < \min\{\lambda^*(x, \xi)/K : L_{t-1}^n \subseteq x \subseteq U_{t-1}^n\} \tag{4.54}$$

and $L_t^n = L_{t-1}^n$ otherwise.

By induction, (4.51) and (4.52) are satisfied, and L^n and U^n are seen to be the maximal respective minimal lower and upper processes with these properties. Hence, if

$$M = \inf\{n > 0 : L^n_{-1} = U^n_{-1}\}$$

denotes the first time a pair of lower and upper processes coalesce, we have that $U^M_{-1} \sim f$. We call M the *coalescence time*.

By induction we have the following *funnelling property*,

$$L^{n-1}_t \subseteq L^n_t \subseteq U^n_t \subseteq U^{n-1}_t, \quad -n \geq t > 0. \tag{4.55}$$

So for $n = 1, 2, 3, \ldots$, we may generate pairs of upper and lower processes as described above, until $L^n_{-1} = U^n_{-1}$, and then return $U^n_{-1} = U^M_{-1}$. It is, however, usually more efficient to use a *doubling scheme*: for $n = 1$ generate Z_{-1}, R_{-1}, U^1, L^1, and for $n = 2, 4, 8, \ldots$, generate $Z_{-n}, R_{-n}, \ldots, Z_{-1-n/2}, R_{-1-n/2}, U^n, L^n$, until $U^n_{-1} = L^n_{-1}$; if N denotes the first such n where this happens, then return $U^N_{-1} \sim f$. This follows from the fact that $U^N_{-1} = U^M_{-1}$, since $M \leq N$ and because of (4.55). As noticed in Propp & Wilson (1996), $N \leq 4M$. As observed in Berthelsen & Møller (2001b), the doubling scheme can be improved slightly by using the scheme $n = m, m+1, m+2, m+4, m+8, \ldots$, where $-m$ denotes the first time a point in Z_{-1} is born.

It may be time consuming to find the maximum and minimum in (4.53) and (4.54) unless the Papangelou conditional intensity satisfies certain *monotonicity properties*: we say that f is *attractive* if $\lambda^*(x, \xi) \leq \lambda^*(y, \xi)$ whenever $x \subset y$; and *repulsive* if $\lambda^*(x, \xi) \geq \lambda^*(y, \xi)$ whenever $x \subset y$; notice that in both cases we easily obtain the maximum and minimum in (4.53) and (4.54). As a matter of fact many point process models satisfy one of these conditions. For instance, the Strauss process (4.31) is repulsive.

Fernández, Ferrari & Garcia (1999) introduce another perfect simulation algorithm based on spatial birth-death processes, but without using upper and lower processes, and with no requirement of monotonicity properties. The algorithm is reviewed and compared with the one using upper and lower processes in Berthelsen & Møller (2001b). In general, if f is attractive or repulsive, the algorithm described in this section is the most efficient.

4.8 Further reading and concluding remarks

Textbooks and review articles on different aspects of spatial point processes include Matérn (1960, 1986), Bartlett (1963, 1964), Kerstan, Matthes & Mecke (1974), Kallenberg (1975), Matheron (1975), Ripley (1977, 1981, 1988), Diggle (1983), Penttinen (1984), Daley & Vere-Jones (1988), Baddeley & Møller (1989), Mecke, Schneider, Stoyan & Weil (1990), Karr (1991), Cressie (1993), Reiss (1993), Stoyan & Stoyan (1994), Stoyan et al. (1995),

Geyer (1999), Møller (1999), van Lieshout (2000), and Ohser & Mücklich (2000).

Examples 1 and 2 concern a single planar pattern; many other examples of simple point processes and marked point processes can be found in the abovementioned books and review papers. Replicated point patterns in 2 and 3 dimensions are discussed in Diggle et al. (1991), Baddeley et al. (1993), and Diggle et al. (2000).

Much of the more classical literature on spatial point processes deal with non-parametric methods for stationary point processes. The focus has changed over the years to exploiting more and more flexible and complex parametric statistical models which are analysed using a Bayesian or likelihood approach by means of MCMC methods as exemplified in this chapter. Most of the literature deal with homogeneous point patterns, but we have also treated the inhomogeneous case in some detail. Jensen & Nielsen (2001) provide a review on some recent developments in this direction.

We expect an increasing need for developing models and tools for analysing inhomogeneous point patterns as new technology such as geographical information systems makes huge amounts of spatial point process data available. Such patterns may exhibit both large-scale aggregation and small-scale inhibition, and simulations may allow to model this using Cox and Markov point processes as building blocks in e.g. a nonparametric Bayesian setting, modifying the approach in Heikkinen & Penttinen (1999) and Berthelsen & Møller (2001a).

We have demonstrated that MCMC methods provide indispensable tools for analysing spatial point process models. Perfect simulation is obviously appealing for many reasons, not at least in connection to simulation-based inference. Its advantages and limitations for statistical inference for spatial point processes are discussed in Berthelsen & Møller (2001a).

Existing codes and *software* may be used in order to save time. Geyer (1999) contains a short description and the address to the code he uses for MCMC likelihood analysis of certain point process models. Adrian Baddeley and Rolf Turner have written spatstat, a contributed library in S-PLUS and R for the statistical analysis of spatial point patterns (available at http://www.maths.uwa.edu.au/~adrian/spatstat/). They provide also references to other spatial statistical software.

However, it is necessary to write new code or modify existing code for many models. The code for a sampler and e.g. MCMC likelihood or Bayesian analysis may be written in C (to speed things up), while a high-level statistical language such as S or R may be used for graphical procedures.

Acknowledgments: Kasper K. Berthelsen and Jakob Gulddahl Rasmussen are acknowledged for their useful comments.

4.9 References

Adler, R. (1981). *The Geometry of Random Fields*, Wiley, New York.

Baddeley, A. & Gill, R. D. (1997). Kaplan-Meier estimators of distance distributions for spatial point processes, *Annals of Statistics* **25**: 263–292.

Baddeley, A. J., Moyeed, R. A., Howard, C. V. & Boyde, A. (1993). Analysis of a three-dimensional point patttern with applications, *Applied Statistics* **42**: 641–668.

Baddeley, A. J. & van Lieshout, M. N. M. (1993). Stochastic geometry models in high-level vision, *in* K. V. Mardia & G. K. Kanji (eds), *Statistics and Images, Advances in Applied Statistics, a supplement to the Journal of Applied Statistics*, Vol. 20, Carfax Publishing, Abingdon, chapter 11, pp. 235–256.

Baddeley, A. J. & van Lieshout, M. N. M. (1995). Area-interaction point processes, *Annals of the Institute of Statistical Mathematics* **46**: 601–619.

Baddeley, A. & Møller, J. (1989). Nearest-neighbour Markov point processes and random sets, *International Statistical Review* **2**: 89–121.

Baddeley, A., Møller, J. & Waagepetersen, R. (2000). Non- and semiparametric estimation of interaction in inhomogeneous point patterns, *Statistica Neerlandica* **54**: 329–350.

Baddeley, A. & Silverman, B. W. (1984). A cautionary example for the use of second-order methods for analysing point patterns, *Biometrics* **40**: 1089–1094.

Baddeley, A. & Turner, R. (2000). Practical maximum pseudolikelihood for spatial point patterns, *Australian and New Zealand Journal of Statistics* **42**: 283–322.

Bartlett, M. S. (1963). The spectral analysis of point processes, *Journal of the Royal Statistical Society Series B* **29**: 264–296.

Bartlett, M. S. (1964). The spectral analysis of two-dimensional point processes, *Biometrika* **51**: 299–311.

Benes, V., Bodlak, K., Møller, J. & Waagepetersen, R. P. (2002). Bayesian analysis of log Gaussian Cox process models for disease mapping, *Technical Report R-02-2001*, Department of Mathematical Sciences, Aalborg University.

Berman, M. & Turner, T. R. (1992). Approximating point process likelihoods with GLIM, *Applied Statistics* **41**: 31–38.

Berthelsen, K. K. & Møller, J. (2001a). Perfect simulation and inference for spatial point processes, *Technical Report R-01-2009*, Department of Mathematical Sciences, Aalborg University. Conditionally accepted for publication in the *Scandinavian Journal of Statistics*.

Berthelsen, K. K. & Møller, J. (2001b). A primer on perfect simulation for spatial point processes, *Technical Report R-01-2026*, Department of Mathematical Sciences, Aalborg University. To appear in *Bulletin of the Brazilian Mathematical Society,* **33**, 2003.

Berthelsen, K. K. & Møller, J. (2002). Spatial jump processes and perfect simulation, *in* K. Mecke & D. Stoyan (eds), *Morphology of Condensed Matter*, Lecture Notes in Physics, Springer-Verlag, Heidelberg. To appear.

Besag, J. (1977a). Some methods of statistical analysis for spatial data, *Bulletin of the Institute of International Statistics* **47**: 77–92.

Besag, J. E. (1974). Spatial interaction and the statistical analysis of lattice systems (with discussion), *Journal of the Royal Statistical Society Series B* **36**: 192–236.

Besag, J. E. (1975). Statistical analysis of non-lattice data, *The Statistician* **24**: 179–195.

Besag, J. E. (1977b). Discussion on the paper by Ripley (1977), *Journal of the Royal Statistical Society Series B* **39**: 193–195.

Besag, J. E. (1994). Discussion on the paper by Grenander and Miller, *Journal of the Royal Statistical Society Series B* **56**: 591–592.

Besag, J., Milne, R. & Zachary, S. (1982). Point process limits of lattice processes, *Journal of Applied Probability* **19**: 210–216.

Breyer, L. A. & Roberts, G. O. (2000). From Metropolis to diffusions: Gibbs states and optimal scaling, *Stochastic Processes and their Applications* **90**: 181–206.

Brix, A. (1999). Generalized gamma measures and shot-noise Cox processes, *Advances in Applied Probability* **31**: 929–953.

Brix, A. & Chadoeuf, J. (2000). Spatio-temporal modeling of weeds and shot-noise G Cox processes. Submitted.

Brix, A. & Kendall, W. S. (2002). Simulation of cluster point processes without edge effects, *Advances in Applied Probability* **34**: 267–280.

Brix, A. & Møller, J. (2001). Space-time multitype log Gaussian Cox processes with a view to modelling weed data, *Scandinavian Journal of Statistics* **28**: 471–488.

Chan, K. S. & Geyer, C. J. (1994). Discussion of the paper 'Markov chains for exploring posterior distributions' by Luke Tierney, *Annals of Statistics* **22**: 1747–1747.

Christensen, O. F., Møller, J. & Waagepetersen, R. P. (2001). Geometric ergodicity of Metropolis-Hastings algorithms for conditional simulation in generalised linear mixed models, *Methodology and Computing in Applied Probability* **3**: 309–327.

Christensen, O. F. & Waagepetersen, R. (2002). Bayesian prediction of spatial count data using generalised linear mixed models, *Biometrics* **58**: 280–286.

Coles, P. & Jones, B. (1991). A lognormal model for the cosmological mass distribution, *Monthly Notices of the Royal Astronomical Society* **248**: 1–13.

Cressie, N. A. C. (1993). *Statistics for Spatial Data*, second edn, Wiley, New York.

Daley, D. J. & Vere-Jones, D. (1988). *An Introduction to the Theory of Point Processes*, Springer-Verlag, New York.

Diggle, P. J. (1983). *Statistical Analysis of Spatial Point Patterns*, Academic Press, London.

Diggle, P. J. (1985). A kernel method for smoothing point process data, *Applied Statistics* **34**: 138–147.

Diggle, P. J., Lange, N. & Beněs, F. (1991). Analysis of variance for replicated spatial point patterns in clinical neuroanatomy, *Journal of the American Statistical Association* **86**: 618–625.

Diggle, P. J., Mateu, L. & Clough, H. E. (2000). A comparison between parametric and non-parametric approaches to the analysis of replicated spatial point patterns, *Advances of Applied Probability* **32**: 331–343.

Fernández, R., Ferrari, P. A. & Garcia, N. L. (1999). Perfect simulation for interacting point processes, loss networks and Ising models. Manuscript.

Fiksel, T. (1984). Estimation of parameterized pair potentials of marked and nonmarked Gibbsian point processes, *Elektronische Informationsverarbeitung und Kypernetik* **20**: 270–278.

Gelfand, A. E. (1996). Model determination using sampling-based methods, *in* W. R. Gilks, S. Richardson & D. J. Spiegelhalter (eds), *Markov chain Monte Carlo in Practice*, Chapman and Hall, London, pp. 145–161.

Gelman, A. & Meng, X.-L. (1998). Simulating normalizing constants: from importance sampling to bridge sampling to path sampling, *Statistical Science* **13**: 163–185.

Georgii, H.-O. (1976). Canonical and grand canonical Gibbs states for continuum systems, *Communications of Mathematical Physics* **48**: 31–51.

Georgii, H.-O. (1988). *Gibbs Measures and Phase Transition*, Walter de Gruyter, Berlin.

Geyer, C. J. (1991). Markov chain Monte Carlo maximum likelihood, *Computing Science and Statistics: Proceedings of the 23rd Symposium on the Interface*, pp. 156–163.

Geyer, C. J. (1994). On the convergence of Monte Carlo maximum likelihood calculations, *Journal of the Royal Society of Statistics Series B* **56**: 261–274.

Geyer, C. J. (1999). Likelihood inference for spatial point processes, *in* O. E. Barndorff-Nielsen, W. S. Kendall & M. N. M. van Lieshout (eds), *Stochastic Geometry: Likelihood and Computation*, Chapman and Hall/CRC, London, Boca Raton, pp. 79–140.

Geyer, C. J. & Møller, J. (1994). Simulation procedures and likelihood inference for spatial point processes, *Scandinavian Journal of Statistics* **21**: 359–373.

Geyer, C. J. & Thompson, E. A. (1992). Constrained Monte Carlo maximum likelihood for dependent data, *Journal of the Royal Society of Statistics Series B* **54**: 657–699.

Goulard, M., Särkkä, A. & Grabarnik, P. (1996). Parameter estimation for marked Gibbs point processes through the maximum pseudo-likelihood method, *Scandinavian Journal of Statistics* **23**: 365–379.

Green, P. J. (1995). Reversible jump MCMC computation and Bayesian model determination, *Biometrika* **82**: 711–732.

Gu, M. G. & Zhu, H.-T. (2001). Maximum likelihood estimation for spatial models by Markov chain Monte Carlo stochastic approximation, *Journal of the Royal Statistical Society Series B* **63**: 339–355.

Häggström, O., van Lieshout, M. N. M. & Møller, J. (1999). Characterization results and Markov chain Monte Carlo algorithms including exact simulation for some spatial point processes, *Bernoulli* **5**: 641–659.

Heikkinen, J. & Arjas, E. (1998). Non-parametric Bayesian estimation of a spatial Poisson intensity, *Scandinavian Journal of Statistics* **25**: 435–450.

Heikkinen, J. & Penttinen, A. (1999). Bayesian smoothing in the estimation of the pair potential function of Gibbs point processes, *Bernoulli* **5**: 1119–1136.

Jensen, E. B. V. & Nielsen, L. S. (2001). A review on inhomogeneous spatial point processes, *in* I. V. Basawa, C. C. Heyde & R. L. Taylor (eds), *Selected Proceedings of the Symposium on Inference for Stochastic Processes*, Vol. 37, IMS Lecture Notes & Monographs Series, Beachwood, Ohio, pp. 297–318.

Jensen, J. L. & Künsch, H. R. (1994). On asymptotic normality of pseudo likelihood estimates for pairwise interaction processes, *Annals of the Institute of Statistical Mathematics* **46**: 475–486.

Jensen, J. L. & Møller, J. (1991). Pseudolikelihood for exponential family models of spatial point processes, *Annals of Applied Probability* **3**: 445–461.

Kallenberg, O. (1975). *Random Measures*, Akadamie-Verlag, Berlin.

Kallenberg, O. (1984). An informal guide to the theory of conditioning in point processes, *International Statistical Review* **52**: 151–164.

Karr, A. F. (1991). *Point Processes and Their Statistical Inference*, Marcel Dekker, New York.

Kelly, F. P. & Ripley, B. D. (1976). A note on Strauss' model for clustering, *Biometrika* **63**: 357–360.

Kendall, W. S. (1998). Perfect simulation for the area-interaction point process, *in* L. Accardi & C. Heyde (eds), *Probability Towards 2000*, Springer, pp. 218–234.

Kendall, W. S. & Møller, J. (2000). Perfect simulation using dominating processes on ordered spaces, with application to locally stable point processes, *Advances in Applied Probability* **32**: 844–865.

Kerscher, M. (2000). Statistical analysis of large-scale structure in the Universe, *in* K. R. Mecke & D. Stoyan (eds), *Statistical Physics and Spatial Statistics*, Lecture Notes in Physics, Springer, Berlin, pp. 36–71.

Kerstan, J., Matthes, K. & Mecke, J. (1974). *Unbegrenzt teilbare Punktprozesse*, Akademie-Verlag, Berlin.

Kingman, J. F. C. (1993). *Poisson Processes*, Clarendon Press, Oxford.

Lieshout, M. N. M. van (2000). *Markov Point Processes and Their Applications*, Imperial College Press, London.

Lieshout, M. N. M. van & Baddeley, A. J. (1996). A nonparametric measure of spatial interaction in point patterns, *Statistica Neerlandica* **50**: 344–361.

Loizeaux, M. A. & McKeague, I. W. (2001). Perfect sampling for posterior landmark distributions with an application to the detection of disease clusters, *in* I. V. Basawa, C. C. Heyde & R. L. Taylor (eds), *Selected Proceedings of the Symposium on Inference for Stochastic Processes*, Vol. 37, IMS Lecture Notes & Monographs Series, Beachwood, Ohio, pp. 321–331.

Lund, J., Penttinen, A. & Rudemo, M. (1999). Bayesian analysis of spatial point patterns from noisy observations. Available at http://www.math.chalmers.se/Stat/ Research/Preprints/.

Lund, J. & Rudemo, M. (2000). Models for point processes observed with noise, *Biometrika* **87**: 235–249.

Lund, J. & Thönnes, E. (2000). Perfect simulation for point processes given noisy observations. Research Report 366, Department of Statistics, University of Warwick.

Mase, S. (1995). Consistency of the maximum pseudo-likelihood estimator of continuous state space Gibbs processes, *Annals of Applied Probability* **5**: 603–612.

Mase, S. (1999). Marked Gibbs processes and asymptotic normality of maximum pseudo-likelihood estimators, *Mathematische Nachrichten* **209**: 151–169.

Mase, S., Møller, J., Stoyan, D., Waagepetersen, R. P. & Döge, G. (2001). Packing densities and simulated tempering for hard core Gibbs point processes, *Annals of the Institute of Statistical Mathematics* **53**: 661–680.

Matérn, B. (1960). Spatial Variation. Meddelanden från Statens Skogforskningsinstitut, Band 49, No. 5.

Matérn, B. (1986). *Spatial Variation*, Lecture Notes in Statistics. Springer-Verlag, Berlin.

Matheron, G. (1975). *Random Sets and Integral Geometry*, Wiley, New York.

Mecke, J. (1967). Stationäre zufällige Maße auf lokalkompakten Abelschen Gruppen, *Zeitschrift für Wahrscheinlichkeitstheorie und verwandte Gebiete* **9**: 36–58.

Mecke, J., Schneider, R. G., Stoyan, D. & Weil, W. R. R. (1990). *Stochastische Geometrie*, Birkhäuser Verlag, Basel.

Metropolis, N., Rosenbluth, A. W., Rosenbluth, M. N., Teller, A. H. & Teller, E. (1953). Equations of state calculations by fast computing machines, *Journal of Chemical Physics* **21**: 1087–1092.

Møller, J. (1989). On the rate of convergence of spatial birth-and-death processes, *Annals of the Institute of Statistical Mathematics* **3**: 565–581.

Møller, J. (1999). Markov chain Monte Carlo and spatial point processes, *in* O. E. Barndorff-Nielsen, W. S. Kendall & M. N. M. van Lieshout (eds), *Stochastic Geometry: Likelihood and Computation*, Monographs on Statistics and Applied Probability 80, Chapman and Hall/CRC, Boca Raton, pp. 141–172.

Møller, J. (2001). A review of perfect simulation in stochastic geometry, *in* I. V. Basawa, C. C. Heyde & R. L. Taylor (eds), *Selected Proceedings of the Symposium on Inference for Stochastic Processes*, Vol. 37, IMS Lecture Notes & Monographs Series, Beachwood, Ohio, pp. 333–355.

Møller, J. (2002a). A comparison of spatial point process models in epidemiological applications, *in* P. J. Green, N. L. Hjort & S. Richardson (eds), *Highly Structured Stochastic Systems*, Oxford University Press, Oxford. To appear.

Møller, J. (2002b). Shot noise Cox processes, *Technical Report R-02-2009*, Department of Mathematical Sciences, Aalborg University.

Møller, J., Syversveen, A. R. & Waagepetersen, R. P. (1998). Log Gaussian Cox processes, *Scandinavian Journal of Statistics* **25**: 451–482.

Møller, J. & Waagepetersen, R. P. (2002). Statistical inference for Cox processes, *in* A. B. Lawson & D. Denison (eds), *Spatial Cluster Modelling*, Chapman and Hall/CRC, Boca Raton.

Møller, J. & Waagepetersen, R. P. (2003). *Statistical Inference and Simulation for Spatial Point Processes*, Chapman and Hall/CRC, Boca Raton. In preparation.

Neyman, J. & Scott, E. L. (1958). Statistical approach to problems of cosmology, *Journal of the Royal Statistical Society Series B* **20**: 1–43.

Nguyen, X. X. & Zessin, H. (1979). Integral and differential characterizations of Gibbs processes, *Mathematische Nachrichten* **88**: 105–115.

Ohser, J. & Mücklich, F. (2000). *Statistical Analysis of Microstructures in Materials Science*, Wiley, New York.

Peebles, P. J. E. (1974). The nature of the distribution of galaxies, *Astronomy and Astrophysics* **32**: 197–202.

Peebles, P. J. E. & Groth, E. J. (1975). Statistical analysis of extragalactic objects. V. Three-point correlation function for the galaxy distribution in the Zwicky catalog, *Astrophysical Journal* **196**: 1–11.

Penttinen, A. (1984). *Modelling Interaction in Spatial Point Patterns: Parameter Estimation by the Maximum Likelihood Method*, Number 7 in Jyväskylä Studies in Computer Science, Economics, and Statistics.

Penttinen, A., Stoyan, D. & Henttonen, H. M. (1992). Marked point processes in forest statistics, *Forest Science* **38**: 806–824.

Preston, C. (1976). *Random Fields*, Lecture Notes in Mathematics, 534. Springer-Verlag, Berlin-Heidelberg.

Preston, C. J. (1977). Spatial birth-and-death processes, *Bulletin of the International Statistical Institute* **46**: 371–391.

Propp, J. G. & Wilson, D. B. (1996). Exact sampling with coupled Markov chains and applications to statistical mechanics, *Random Structures and Algorithms* **9**: 223–252.

Quine, M. P. & Watson, D. F. (1984). Radial simulation of n-dimensional Poisson processes, *Journal of Applied Probability* **21**: 548–557.

Rathbun, S. L. (1996). Estimation of Poisson intensity using partially observed concomitant variables, *Biometrics* **52**: 226–242.

Reiss, R.-D. (1993). *A Course on Point Processes*, Springer Verlag, New York.

Ripley, B. D. (1977). Modelling spatial patterns (with discussion), *Journal of the Royal Statistical Society Series B* **39**: 172–212.

Ripley, B. D. (1979). Simulating spatial patterns: dependent samples from a multivariate density. Algorithm AS 137, *Applied Statistics* **28**: 109–112.

Ripley, B. D. (1981). *Spatial Statistics*, Wiley, New York.

Ripley, B. D. (1988). *Statistical Inference for Spatial Processes*, Cambridge University Press, Cambridge.

Ripley, B. D. & Kelly, F. P. (1977). Markov point processes, *Journal of the London Mathematical Society* **15**: 188–192.

Roberts, G. O., Gelman, A. & Gilks, W. R. (1997). Weak convergence and optimal scaling of random walk Metropolis algorithms, *Annals of Applied Probability* **7**: 110–120.

Roberts, G. O. & Rosenthal, J. S. (1997). Geometric ergodicity and hybrid Markov chains, *Electronic Communications in Probability* **2**: 13–25.

Roberts, G. O. & Rosenthal, J. S. (1998). Optimal scaling of discrete approximations to Langevin diffusions, *Journal of the Royal Statistical Society Series B* **60**: 255–268.

Roberts, G. O. & Tweedie, R. L. (1996). Exponential convergence of Langevin diffusions and their discrete approximations, *Bernoulli* **2**: 341–363.

Rossky, P. J., Doll, J. D. & Friedman, H. L. (1978). Brownian dynamics as smart Monte Carlo simulation, *Journal of Chemical Physics* **69**: 4628–4633.

Ruelle, D. (1969). *Statistical Mechanics: Rigorous Results*, W.A. Benjamin, Reading, Massachusetts.

Schladitz, K. & Baddeley, A. J. (2000). A third-order point process characteristic, *Scandinavian Journal of Statistics* **27**: 657–671.

Schlather, M. (2001). On the second-order characteristics of marked point processes, *Bernoulli* **7**: 99–117.

Stoyan, D., Kendall, W. S. & Mecke, J. (1995). *Stochastic Geometry and Its Applications*, second edn, Wiley, Chichester.

Stoyan, D. & Stoyan, H. (1994). *Fractals, Random Shapes and Point Fields*, Wiley, Chichester.

Stoyan, D. & Stoyan, H. (2000). Improving ratio estimators of second order point process characteristics, *Scandinavian Journal of Statistics* **27**: 641–656.

Strauss, D. J. (1975). A model for clustering, *Biometrika* **63**: 467–475.

Thönnes, E. (1999). Perfect simulation of some point processes for the impatient user, *Advances in Applied Probability* **31**: 69–87.

Waagepetersen, R. & Sorensen, S. (2001). A tutorial on reversible jump MCMC with a view toward applications in QTL-mapping, *International Statistical Review* **69**(1): 49–61.

Wolpert, R. L. & Ickstadt, K. (1998). Poisson/gamma random field models for spatial statistics, *Biometrika* **85**: 251–267.

Wood, A. T. A. & Chan, G. (1994). Simulation of stationary Gaussian processes in $[0,1]^d$, *Journal of Computational and Graphical Statistics* **3**: 409–432.

Index

Lecture Notes in Statistics

For information about Volumes 1 to 120, please contact Springer-Verlag

121: Constantine Gatsonis, James S. Hodges, Robert E. Kass, Robert McCulloch, Peter Rossi, and Nozer D. Singpurwalla (Editors), Case Studies in Bayesian Statistics, Volume III. xvi, 487 pp., 1997.

122: Timothy G. Gregoire, David R. Brillinger, Peter J. Diggle, Estelle Russek-Cohen, William G. Warren, and Russell D. Wolfinger (Editors), Modeling Longitudinal and Spatially Correlated Data. x, 402 pp., 1997.

123: D. Y. Lin and T. R. Fleming (Editors), Proceedings of the First Seattle Symposium in Biostatistics: Survival Analysis. xiii, 308 pp., 1997.

124: Christine H. Müller, Robust Planning and Analysis of Experiments. x, 234 pp., 1997.

125: Valerii V. Fedorov and Peter Hackl, Model-Oriented Design of Experiments. viii, 117 pp., 1997.

126: Geert Verbeke and Geert Molenberghs, Linear Mixed Models in Practice: A SAS-Oriented Approach. xiii, 306 pp., 1997.

127: Harald Niederreiter, Peter Hellekalek, Gerhard Larcher, and Peter Zinterhof (Editors), Monte Carlo and Quasi-Monte Carlo Methods 1996. xii, 448 pp., 1997.

128: L. Accardi and C.C. Heyde (Editors), Probability Towards 2000. x, 356 pp., 1998.

129: Wolfgang Härdle, Gerard Kerkyacharian, Dominique Picard, and Alexander Tsybakov, Wavelets, Approximation, and Statistical Applications. xvi, 265 pp., 1998.

130: Bo-Cheng Wei, Exponential Family Nonlinear Models. ix, 240 pp., 1998.

131: Joel L. Horowitz, Semiparametric Methods in Econometrics. ix, 204 pp., 1998.

132: Douglas Nychka, Walter W. Piegorsch, and Lawrence H. Cox (Editors), Case Studies in Environmental Statistics. viii, 200 pp., 1998.

133: Dipak Dey, Peter Müller, and Debajyoti Sinha (Editors), Practical Nonparametric and Semiparametric Bayesian Statistics. xv, 408 pp., 1998.

134: Yu. A. Kutoyants, Statistical Inference For Spatial Poisson Processes. vii, 284 pp., 1998.

135: Christian P. Robert, Discretization and MCMC Convergence Assessment. x, 192 pp., 1998.

136: Gregory C. Reinsel, Raja P. Velu, Multivariate Reduced-Rank Regression. xiii, 272 pp., 1998.

137: V. Seshadri, The Inverse Gaussian Distribution: Statistical Theory and Applications. xii, 360 pp., 1998.

138: Peter Hellekalek and Gerhard Larcher (Editors), Random and Quasi-Random Point Sets. xi, 352 pp., 1998.

139: Roger B. Nelsen, An Introduction to Copulas. xi, 232 pp., 1999.

140: Constantine Gatsonis, Robert E. Kass, Bradley Carlin, Alicia Carriquiry, Andrew Gelman, Isabella Verdinelli, and Mike West (Editors), Case Studies in Bayesian Statistics, Volume IV. xvi, 456 pp., 1999.

141: Peter Müller and Brani Vidakovic (Editors), Bayesian Inference in Wavelet Based Models. xiii, 394 pp., 1999.

142: György Terdik, Bilinear Stochastic Models and Related Problems of Nonlinear Time Series Analysis: A Frequency Domain Approach. xi, 258 pp., 1999.

143: Russell Barton, Graphical Methods for the Design of Experiments. x, 208 pp., 1999.

144: L. Mark Berliner, Douglas Nychka, and Timothy Hoar (Editors), Case Studies in Statistics and the Atmospheric Sciences. x, 208 pp., 2000.

145: James H. Matis and Thomas R. Kiffe, Stochastic Population Models. viii, 220 pp., 2000.

146: Wim Schoutens, Stochastic Processes and Orthogonal Polynomials. xiv, 163 pp., 2000.

147: Jürgen Franke, Wolfgang Härdle, and Gerhard Stahl, Measuring Risk in Complex Stochastic Systems. xvi, 272 pp., 2000.

148: S.E. Ahmed and Nancy Reid, Empirical Bayes and Likelihood Inference. x, 200 pp., 2000.

149: D. Bosq, Linear Processes in Function Spaces: Theory and Applications. xv, 296 pp., 2000.

150: Tadeusz Caliński and Sanpei Kageyama, Block Designs: A Randomization Approach, Volume I: Analysis. ix, 313 pp., 2000.

151: Håkan Andersson and Tom Britton, Stochastic Epidemic Models and Their Statistical Analysis. ix, 152 pp., 2000.

152: David Ríos Insua and Fabrizio Ruggeri, Robust Bayesian Analysis. xiii, 435 pp., 2000.

153: Parimal Mukhopadhyay, Topics in Survey Sampling. x, 303 pp., 2000.

154: Regina Kaiser and Agustín Maravall, Measuring Business Cycles in Economic Time Series. vi, 190 pp., 2000.

155: Leon Willenborg and Ton de Waal, Elements of Statistical Disclosure Control. xvii, 289 pp., 2000.

156: Gordon Willmot and X. Sheldon Lin, Lundberg Approximations for Compound Distributions with Insurance Applications. xi, 272 pp., 2000.

157: Anne Boomsma, Marijtje A.J. van Duijn, and Tom A.B. Snijders (Editors), Essays on Item Response Theory. xv, 448 pp., 2000.

158: Dominique Ladiray and Benoît Quenneville, Seasonal Adjustment with the X-11 Method. xxii, 220 pp., 2001.

159: Marc Moore (Editor), Spatial Statistics: Methodological Aspects and Some Applications. xvi, 282 pp., 2001.

160: Tomasz Rychlik, Projecting Statistical Functionals. viii, 184 pp., 2001.

161: Maarten Jansen, Noise Reduction by Wavelet Thresholding. xxii, 224 pp., 2001.

162: Constantine Gatsonis, Bradley Carlin, Alicia Carriquiry, Andrew Gelman, Robert E. Kass Isabella Verdinelli, and Mike West (Editors), Case Studies in Bayesian Statistics, Volume V. xiv, 448 pp., 2001.

163: Erkki P. Liski, Nripes K. Mandal, Kirti R. Shah, and Bikas K. Sinha, Topics in Optimal Design. xii, 164 pp., 2002.

164: Peter Goos, The Optimal Design of Blocked and Split-Plot Experiments. xiv, 244 pp., 2002.

165: Karl Mosler, Multivariate Dispersion, Central Regions and Depth: The Lift Zonoid Approach. xii, 280 pp., 2002.

166: Hira L. Koul, Weighted Empirical Processes in Dynamic Nonlinear Models, Second Edition. xiii, 425 pp., 2002.

167: Constantine Gatsonis, Alicia Carriquiry, Andrew Gelman, David Higdon, Robert E. Kass, Donna Pauler, and Isabella Verdinelli (Editors), Case Studies in Bayesian Statistics, Volume VI. xiv, 376 pp., 2002.

168: Susanne Rässler, Statistical Matching: A Frequentist Theory, Practical Applications, nd Alternative Bayesian Approaches. xviii, 238 pp., 2002.

169: Yu. I. Ingster and Irina A. Suslina, Nonparametric Goodness-of-Fit Testing Under Gaussian Models. xiv, 453 pp., 2003.

170: Tadeusz Caliński and Sanpei Kageyama, Block Designs: A Randomization Approach, Volume II: Design. xii, 351 pp., 2003.

171: David D. Denison, Mark H. Hansen, Christopher C. Holmes, Bani Mallick, Bin Yu (Editors) Nonlinear Estimation and Classification. viii, 474 pp., 2003.

172: Sneh Gulati, William J. Padgett, Parametric and Nonparametric Inference from Record-Breaking Data. ix, 112 pp., 2003.

173: Jesper Møller (Editor), Spatial Statistics and Computational Methods. xiv, 202 pp., 2003.

9 780387 001364